W9-AZW-891

Mastering Tables

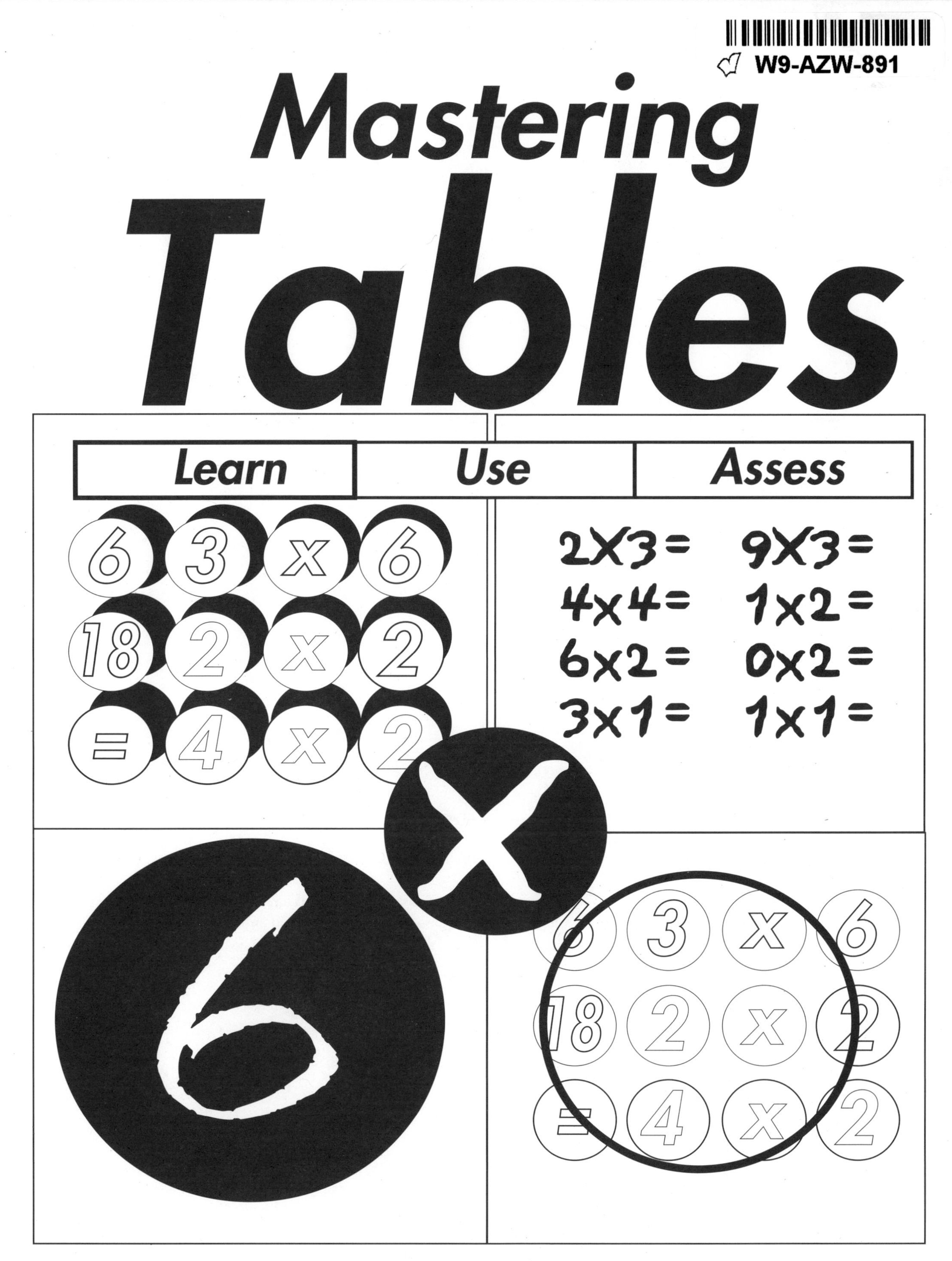

Written by Murray Brennan

Published by World Teachers Press®
www.worldteacherspress.com

Published with the permission of R.I.C. Publications Pty. Ltd.

Copyright © 2005 by Didax, Inc., Rowley, MA 01969. All rights reserved.

First published by R.I.C. Publications Pty. Ltd., Perth, Western Australia. Revised by Didax Educational Resources.

Limited reproduction permission: The publisher grants permission to individual teachers who have purchased this book to reproduce the blackline masters as needed for use with their own students. Reproduction for an entire school or school district or for commercial use is prohibited.

Printed in the United States of America.

This book is printed on recycled paper.

Order Number 2-5235
ISBN 978-1-58324-199-8

C D E F G 12 11 10 09 08

395 Main Street
Rowley, MA 01969
www.didax.com

Foreword

Mastering Tables *is a comprehensive program for teachers to help students in learning, practicing, assessing, and extending their knowledge and recall of multiplication tables through challenging, fun activities.*

Research shows that student's instant recall of basic number facts will only progress from short-term memory (easily forgotten) to the long-term memory through constant practice and reinforcement of the same table facts. Mastering Tables *provides you with a variety of activities and techniques to help students achieve instant recall and understanding of related times table facts.*

Students will enjoy the self-competition aspect of these activities while at the same time reinforcing their knowledge of essential facts.

An ideal mathematical support program, Mastering Tables *provides a framework to:*

- *encourage and develop mental calculation skills*
- *develop problem-solving strategies and skills*
- *develop and maintain speed of recall*
- *provide support to the overall daily mathematics program*

Contents

©World Teachers Press® www.worldteacherspress.com

Teacher's notes

Mastering Tables *provides a variety of activities to facilitate remediation, practice, and extension of basic times table facts.*

Covering the basic tables, 2 – 10, Mastering Tables *also extends students to consider the 12, 15, 20, 25, and 50 times tables. Derivatives for each table are covered throughout the related activities. For example, as the three times table facts and patterns are studied so to are the 30 times table facts; four times table studies the 40 times table and so on.*

You can use Mastering Tables *for formal classroom lessons, review, and reinforcement, extension, or homework activities.*

Calculators may be a useful mathematical tool for students when tackling more difficult stages of these multiplication tables.

The activities are divided into four main learning areas.

(A) Learning your tables

Set A or B can be used in separate lessons to reinforce simple table facts.

For those students having difficulty or wanting to improve their accuracy scores, this activity page can be used for homework or review activities.

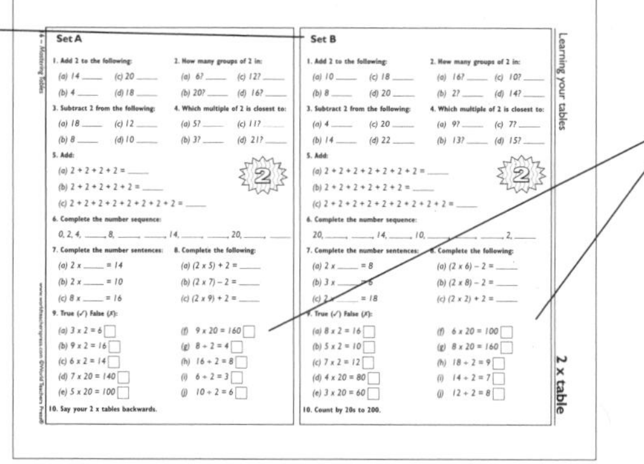

Provides two sets of mathematical activities using the four basic operations to promote the learning of each table. The emphasis is on understanding rather than rote learning.

(B) Using your tables

Recording the basic times table answers to assist in recall and application in related activities.

Activities enabling students to apply tables knowledge to solve everyday mathematical problems.

Simple number charts or fun games like "Tables Battleship" or "Chutes 'n' Ladders" to reinforce student learning in an enjoyable, relevant way. (See page 6 for game instructions.)

Recording and learning the derivatives of the times table.

(C) Assessing your tables

Space provided for students to record their score and times as they race against themselves and the clock. (See instructions for time recordings on page 5.)

Students will want to learn and improve their times tables as they race against themselves in these speed and accuracy activities for each multiplication table.

Two sets of 40 mental calculations to use as separate lessons for reinforcement and assessment of multiplication tables.

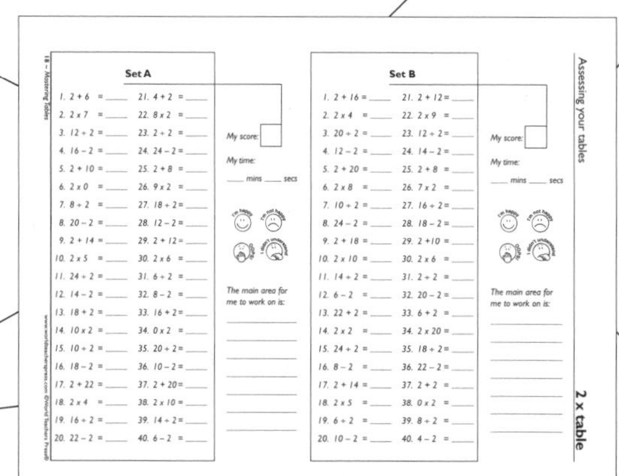

For each multiplication table there is a choice of two set activities (A or B) for the students to complete as you direct.

Self-evaluation, where students can select the facial expression that best describes their feelings about their results. Students can also comment on their successes or where they think they could improve.

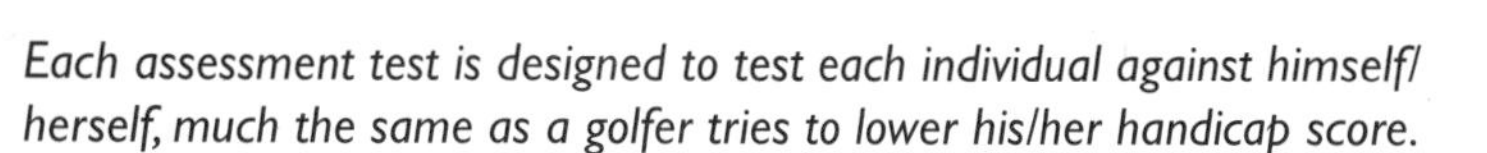

Each assessment test is designed to test each individual against himself/herself, much the same as a golfer tries to lower his/her handicap score.

www.worldteacherspress.com ©World Teachers Press®

Teacher's notes

Timing

- *Each student begins with the maximum time allowance of 3 minutes.*
- *Students mark their answers (provided on pages 63 – 68) to get a score and color their progress on the "Student progress graph" (pages 7 – 8).*
- *The students need to work out their time allowance for the next day/lesson. Those with a perfect score (40) will reduce their times to 2 minutes. Those who did not obtain a perfect score remain on 3 minutes.*
- *Each time a student scores a perfect score, his/her time is further reduced by 30 seconds, making each test more challenging. Students with perfect scores move from 3 minutes to 2 minutes to 90 seconds and finally to 60 seconds.*
- *Those students not scoring 40 will remain on the same time frame until they are directed by you or until they score 40 correct answers.*

Student progress graphs have been included on pages 7 – 8.

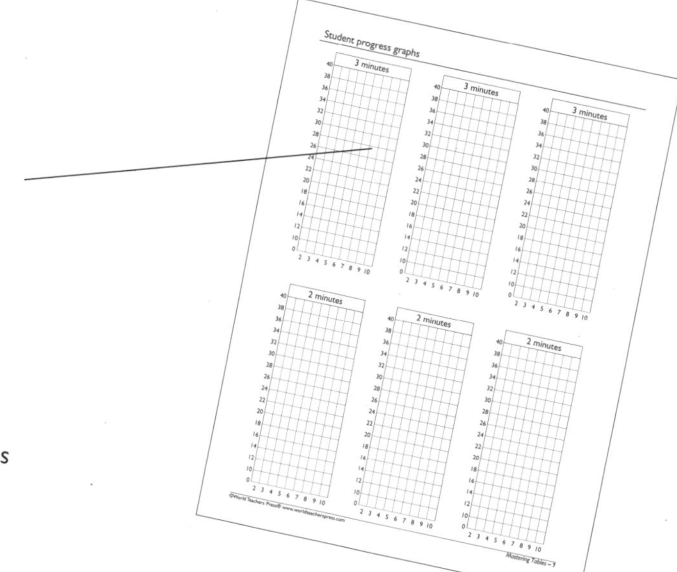

(D) Reviewing your tables

Two sets of 40 problems to review and assess a combination of multiplication tables.

Review sheets use the four basic operations and the combination of two different multiplication tables.

Self-evaluation, where students can select the facial expression that best describes their feelings in relation to their results. Students can also comment on their successes or where they think they could improve.

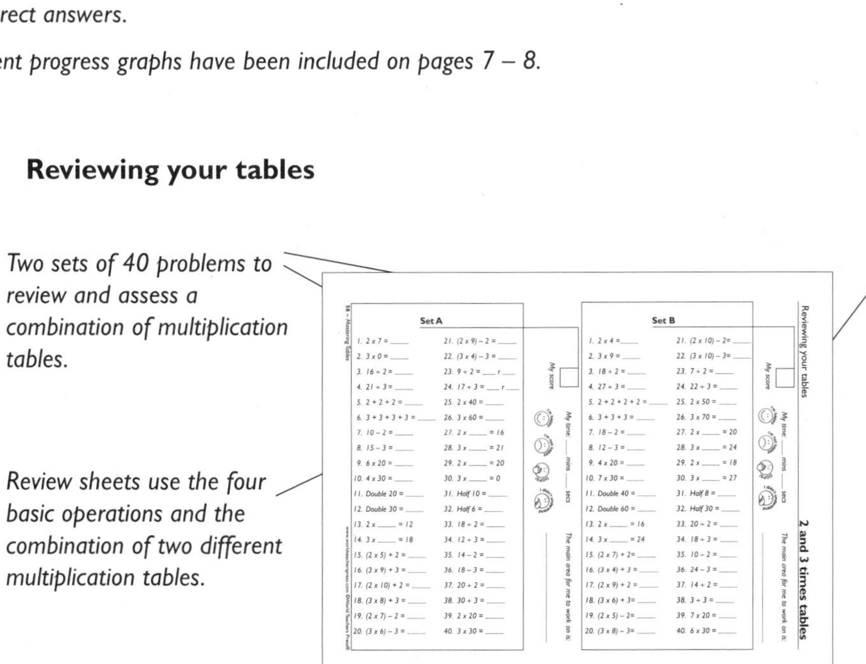

Reviewing your tables

2 and 3 times tables

Set A

1. 2 x 7 = ____
2. 3 x 0 = ____
3. 16 ÷ 2 = ____
4. 21 ÷ 3 = ____
5. 2 + 2 + 2 = ____
6. 3 + 3 + 3 + 3 = ____
7. 10 – 2 = ____
8. 15 – 3 = ____
9. 6 x 20 = ____
10. 4 x 30 = ____
11. Double 20 = ____
12. Double 30 = ____
13. 2 x ____ = 12
14. 3 x ____ = 18
15. (2 x 5) + 2 = ____
16. (3 x 9) + 3 = ____
17. (2 x 10) + 2 = ____
18. (3 x 8) + 3 = ____
19. (2 x 7) – 2 = ____
20. (3 x 6) – 3 = ____
21. (2 x 9) – 2 = ____
22. (3 x 4) – 3 = ____
23. 9 ÷ 2 = ____ r ____
24. 17 ÷ 3 = ____ r ____
25. 2 x 40 = ____
26. 3 x 60 = ____
27. 2 x ____ = 16
28. 3 x ____ = 21
29. 2 x ____ = 20
30. 3 x ____ = 0
31. Half 10 = ____
32. Half 6 = ____
33. 18 ÷ 2 = ____
34. 12 ÷ 3 = ____
35. 14 – 2 = ____
36. 18 – 3 = ____
37. 20 ÷ 2 = ____
38. 30 ÷ 3 = ____
39. 2 x 20 = ____
40. 3 x 30 = ____

My score

My time: ____ mins ____ secs

The main area for me to work on is:

Set B

1. 2 x 4 = ____
2. 3 x 9 = ____
3. 18 ÷ 2 = ____
4. 27 ÷ 3 = ____
5. 2 + 2 + 2 + 2 = ____
6. 3 + 3 + 3 = ____
7. 18 – 2 = ____
8. 12 – 3 = ____
9. 4 x 20 = ____
10. 7 x 30 = ____
11. Double 40 = ____
12. Double 60 = ____
13. 2 x ____ = 16
14. 3 x ____ = 24
15. (2 x 7) + 2= ____
16. (3 x 4) + 3 = ____
17. (2 x 9) + 2 = ____
18. (3 x 6) + 3= ____
19. (2 x 5) – 2= ____
20. (3 x 8) – 3= ____
21. (2 x 10) – 2= ____
22. (3 x 10) – 3= ____
23. 7 ÷ 2 = ____
24. 22 ÷ 3 = ____
25. 2 x 50 = ____
26. 3 x 70 = ____
27. 2 x ____ = 20
28. 3 x ____ = 24
29. 2 x ____ = 18
30. 3 x ____ = 27
31. Half 8 = ____
32. Half 30 = ____
33. 20 ÷ 2 = ____
34. 18 ÷ 3 = ____
35. 10 – 2 = ____
36. 24 – 3 = ____
37. 14 ÷ 2 = ____
38. 3 ÷ 3 = ____
39. 7 x 20 = ____
40. 6 x 30 = ____

My score

My time: ____ mins ____ secs

The main area for me to work on is:

58 ~ Mastering Tables

www.worldteacherspress.com ©World Teachers Press®

Marking

Marking is simple. Answers are provided. There are several ways you might like to arrange for answers to be marked. Remember, mental math sessions should be brief and to the point, so whichever method you choose needs to be efficient while identifying potential problem areas.

Methods could include:

- *whole-class: you call the answers to students who either self-check or partner-check.*
- *student checking: students use answer sheets provided to either self-check or partner-check.*
- *collection: you collect the worksheets to mark individual student work. This can be a useful process to check accuracy of other methods used.*

Games

(a) **Times Battleship** *(found on pages 44, 47, 50, 53, and 56)*

A game for 2 players.

Each player has a squared grid with coordinates A, B, C, D and 1, 2, 3, 4. Each player fills in seven multiples of the designated table. The remaining squares are filled with the word "miss." Players then take turns locating the other player's "ships" (multiples) by calling coordinates; e.g. B4.

For example (x 6 table grid shown)

	A	B	C	D
4	12	~~48~~	miss	54
3	miss	42	miss	miss
2	6	miss	miss	18
1	miss	miss	30	miss

If a player calls "B4," the other player calls "ship 48." If the first player calls out the correct times table for 48 (i.e. 6 x 8) then that "ship" is "sunk" and crossed out. The winner is the first player to sink all 7 "ships" belonging to the other player.

(b) Chutes 'n' Ladders *(found on pages 29, 32, 35, 38, and 41)*

A game for 2 players.

Color each number on the gameboard which is a multiple of the assigned times table; e.g. 7 times table—7, 14, 21, 28 .... Draw 4 ladders from colored multiples to higher numbers. Draw 4 chutes from colored multiples and leading down to lower numbers. Students will need a die and two small counters. To start from the number 1, students will need to throw a 6 on the die. If a counter lands on a ladder it is moved up to the number at the top of the ladder. If a counter lands on a chute, it is moved down to the number at the base of the chute. The first player to reach 100 wins the game.

(c) Tables Bingo *(see page 14 for templates.)*

You select a times table grid that students will play in. Fill in the squares of that bingo grid with multiples of the number in the center. (Multiples are numbers that a given number will divide into exactly. Multiples of 3, for example, are 3, 6, 9, 12, 15, etc.) Call out a multiplication problem from the allotted table; e.g. 3 x 5. If the answer is on a student's grid he/she crosses it off. Students call out "Bingo" when a line (across, down, or diagonally) or all numbers are crossed. (To be determined by you before the game commences.)

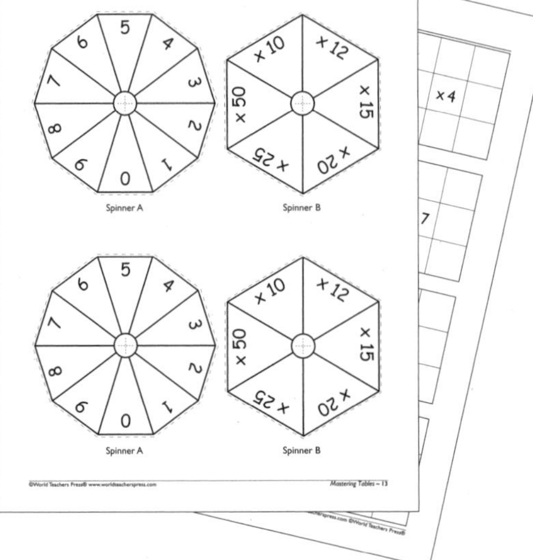

(d) Tables Spinners *(see page 13 for templates.)*

Copy spinners onto thick card. Cut and color the spinners and use a pencil or similar object to poke through the center. You or student leaders, in small groups, can choose a table orally; e.g. 5 times table. Students take turns to spin the appropriate spinner to match the selected table. Students record the multiplication answers when the number the spinner lands on is multiplied by the chosen table. As students get more competent with the game, perhaps time limits could be set to test and improve recall.

Merit certificates have been included for positive reinforcement of students' progress in Mastering Tables. (See page 15)

Finding Multiples (see pages 11 – 12)

Easy to use, timesaving multiple grids to increase students' knowledge of multiples related to the multiplication tables. Multiples are numbers that a given number will divide into exactly; e.g. multiples of 5 are 5, 10, 15, 20, etc. Discuss the patterns found.

www.worldteacherspress.com ©World Teachers Press®

Student progress graphs

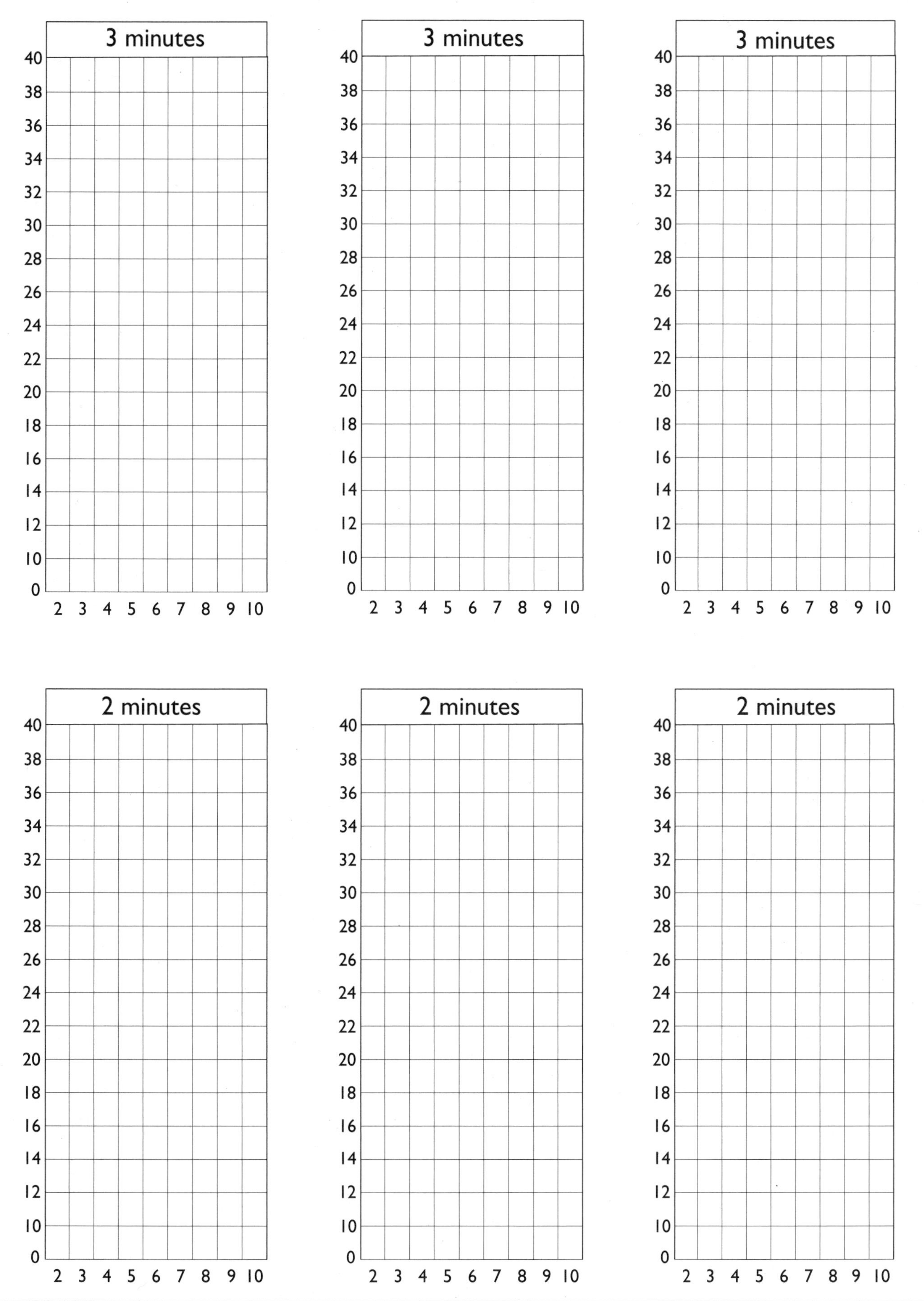

Student progress graphs

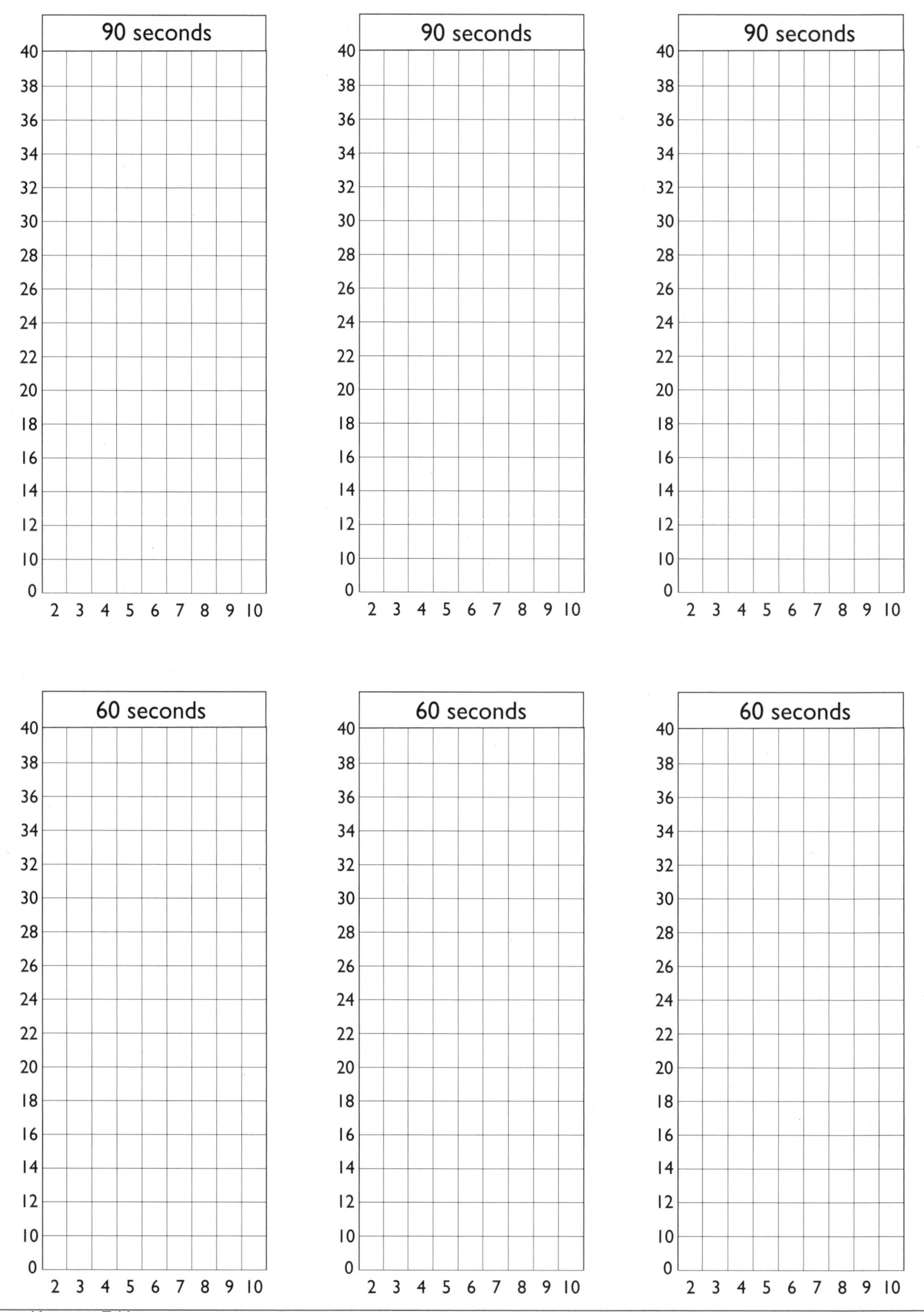

www.worldteacherspress.com ©World Teachers Press®

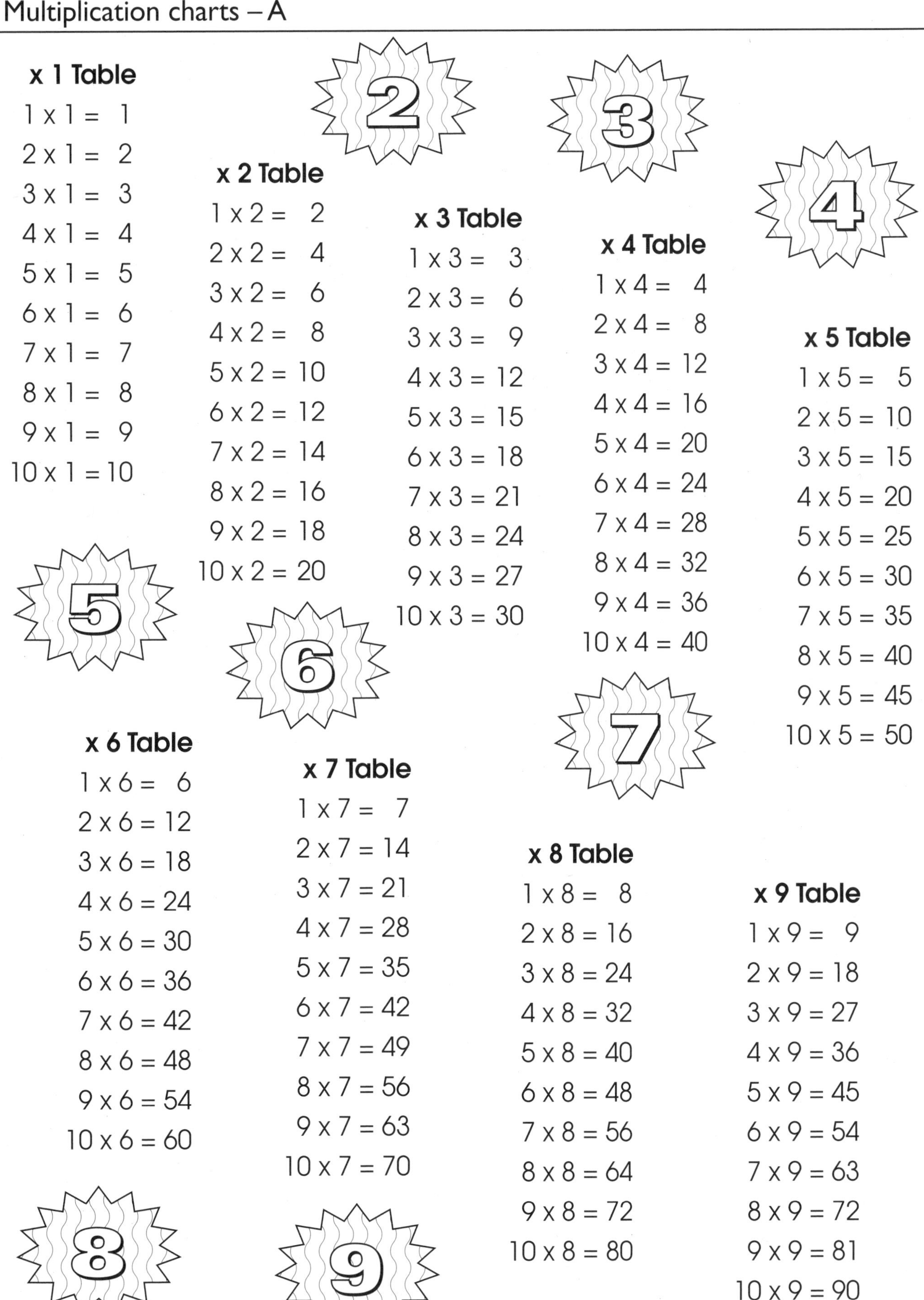

x 1 Table
1 x 1 = 1
2 x 1 = 2
3 x 1 = 3
4 x 1 = 4
5 x 1 = 5
6 x 1 = 6
7 x 1 = 7
8 x 1 = 8
9 x 1 = 9
10 x 1 = 10

x 2 Table
1 x 2 = 2
2 x 2 = 4
3 x 2 = 6
4 x 2 = 8
5 x 2 = 10
6 x 2 = 12
7 x 2 = 14
8 x 2 = 16
9 x 2 = 18
10 x 2 = 20

x 3 Table
1 x 3 = 3
2 x 3 = 6
3 x 3 = 9
4 x 3 = 12
5 x 3 = 15
6 x 3 = 18
7 x 3 = 21
8 x 3 = 24
9 x 3 = 27
10 x 3 = 30

x 4 Table
1 x 4 = 4
2 x 4 = 8
3 x 4 = 12
4 x 4 = 16
5 x 4 = 20
6 x 4 = 24
7 x 4 = 28
8 x 4 = 32
9 x 4 = 36
10 x 4 = 40

x 5 Table
1 x 5 = 5
2 x 5 = 10
3 x 5 = 15
4 x 5 = 20
5 x 5 = 25
6 x 5 = 30
7 x 5 = 35
8 x 5 = 40
9 x 5 = 45
10 x 5 = 50

x 6 Table
1 x 6 = 6
2 x 6 = 12
3 x 6 = 18
4 x 6 = 24
5 x 6 = 30
6 x 6 = 36
7 x 6 = 42
8 x 6 = 48
9 x 6 = 54
10 x 6 = 60

x 7 Table
1 x 7 = 7
2 x 7 = 14
3 x 7 = 21
4 x 7 = 28
5 x 7 = 35
6 x 7 = 42
7 x 7 = 49
8 x 7 = 56
9 x 7 = 63
10 x 7 = 70

x 8 Table
1 x 8 = 8
2 x 8 = 16
3 x 8 = 24
4 x 8 = 32
5 x 8 = 40
6 x 8 = 48
7 x 8 = 56
8 x 8 = 64
9 x 8 = 72
10 x 8 = 80

x 9 Table
1 x 9 = 9
2 x 9 = 18
3 x 9 = 27
4 x 9 = 36
5 x 9 = 45
6 x 9 = 54
7 x 9 = 63
8 x 9 = 72
9 x 9 = 81
10 x 9 = 90

©World Teachers Press® www.worldteacherspress.com

x 10 Table

1 x 10 = 10
2 x 10 = 20
3 x 10 = 30
4 x 10 = 40
5 x 10 = 50
6 x 10 = 60
7 x 10 = 70
8 x 10 = 80
9 x 10 = 90
10 x 10 = 100

x 12 Table

1 x 12 = 12
2 x 12 = 24
3 x 12 = 36
4 x 12 = 48
5 x 12 = 60
6 x 12 = 72
7 x 12 = 84
8 x 12 = 96
9 x 12 = 108
10 x 12 = 120

x 15 Table

1 x 15 = 15
2 x 15 = 30
3 x 15 = 45
4 x 15 = 60
5 x 15 = 75
6 x 15 = 90
7 x 15 = 105
8 x 15 = 120
9 x 15 = 135
10 x 15 = 150

x 20 Table

1 x 20 = 20
2 x 20 = 40
3 x 20 = 60
4 x 20 = 80
5 x 20 = 100
6 x 20 = 120
7 x 20 = 140
8 x 20 = 160
9 x 20 = 180
10 x 20 = 200

x 25 Table

1 x 25 = 25
2 x 25 = 50
3 x 25 = 75
4 x 25 = 100
5 x 25 = 125
6 x 25 = 150
7 x 25 = 175
8 x 25 = 200
9 x 25 = 225
10 x 25 = 250

x 50 Table

1 x 50 = 50
2 x 50 = 100
3 x 50 = 150
4 x 50 = 200
5 x 50 = 250
6 x 50 = 300
7 x 50 = 350
8 x 50 = 400
9 x 50 = 450
10 x 50 = 500

www.worldteacherspress.com ©World Teachers Press®

Finding multiples

Mark the multiples of 2, 3, 4, 5, 6, and 7 on the 1 – 100 grids provided.

1	2	3	4	5	6	7	8	9	10
11	12	13	14	15	16	17	18	19	20
21	22	23	24	25	26	27	28	29	30
31	32	33	34	35	36	37	38	39	40
41	42	43	44	45	46	47	48	49	50
51	52	53	54	55	56	57	58	59	60
61	62	63	64	65	66	67	68	69	70
71	72	73	74	75	76	77	78	79	80
81	82	83	84	85	86	87	88	89	90
91	92	93	94	95	96	97	98	99	100

Multiples of 2

1	2	3	4	5	6	7	8	9	10
11	12	13	14	15	16	17	18	19	20
21	22	23	24	25	26	27	28	29	30
31	32	33	34	35	36	37	38	39	40
41	42	43	44	45	46	47	48	49	50
51	52	53	54	55	56	57	58	59	60
61	62	63	64	65	66	67	68	69	70
71	72	73	74	75	76	77	78	79	80
81	82	83	84	85	86	87	88	89	90
91	92	93	94	95	96	97	98	99	100

Multiples of 3

What is a multiple?

The multiples of 2 would be 2, 4, 6, 8, 10, 12, etc.

1	2	3	4	5	6	7	8	9	10
11	12	13	14	15	16	17	18	19	20
21	22	23	24	25	26	27	28	29	30
31	32	33	34	35	36	37	38	39	40
41	42	43	44	45	46	47	48	49	50
51	52	53	54	55	56	57	58	59	60
61	62	63	64	65	66	67	68	69	70
71	72	73	74	75	76	77	78	79	80
81	82	83	84	85	86	87	88	89	90
91	92	93	94	95	96	97	98	99	100

Multiples of 4

1	2	3	4	5	6	7	8	9	10
11	12	13	14	15	16	17	18	19	20
21	22	23	24	25	26	27	28	29	30
31	32	33	34	35	36	37	38	39	40
41	42	43	44	45	46	47	48	49	50
51	52	53	54	55	56	57	58	59	60
61	62	63	64	65	66	67	68	69	70
71	72	73	74	75	76	77	78	79	80
81	82	83	84	85	86	87	88	89	90
91	92	93	94	95	96	97	98	99	100

Multiples of 5

What happens if you use a different number grid?

1	2	3	4	5	6	7	8	9	10
11	12	13	14	15	16	17	18	19	20
21	22	23	24	25	26	27	28	29	30
31	32	33	34	35	36	37	38	39	40
41	42	43	44	45	46	47	48	49	50
51	52	53	54	55	56	57	58	59	60
61	62	63	64	65	66	67	68	69	70
71	72	73	74	75	76	77	78	79	80
81	82	83	84	85	86	87	88	89	90
91	92	93	94	95	96	97	98	99	100

Multiples of 6

1	2	3	4	5	6	7	8	9	10
11	12	13	14	15	16	17	18	19	20
21	22	23	24	25	26	27	28	29	30
31	32	33	34	35	36	37	38	39	40
41	42	43	44	45	46	47	48	49	50
51	52	53	54	55	56	57	58	59	60
61	62	63	64	65	66	67	68	69	70
71	72	73	74	75	76	77	78	79	80
81	82	83	84	85	86	87	88	89	90
91	92	93	94	95	96	97	98	99	100

Multiples of 7

Try to describe each pattern.

Finding multiples

Mark the multiples of 8, 9, 10, 12, and 15 on the 1 – 100 grids provided.

What is a multiple?

1	2	3	4	5	6	7	8	9	10
11	12	13	14	15	16	17	18	19	20
21	22	23	24	25	26	27	28	29	30
31	32	33	34	35	36	37	38	39	40
41	42	43	44	45	46	47	48	49	50
51	52	53	54	55	56	57	58	59	60
61	62	63	64	65	66	67	68	69	70
71	72	73	74	75	76	77	78	79	80
81	82	83	84	85	86	87	88	89	90
91	92	93	94	95	96	97	98	99	100

Multiples of 8

1	2	3	4	5	6	7	8	9	10
11	12	13	14	15	16	17	18	19	20
21	22	23	24	25	26	27	28	29	30
31	32	33	34	35	36	37	38	39	40
41	42	43	44	45	46	47	48	49	50
51	52	53	54	55	56	57	58	59	60
61	62	63	64	65	66	67	68	69	70
71	72	73	74	75	76	77	78	79	80
81	82	83	84	85	86	87	88	89	90
91	92	93	94	95	96	97	98	99	100

Multiples of 9

The multiples of 8 would be 8, 16, 24, 32, etc.

1	2	3	4	5	6	7	8	9	10
11	12	13	14	15	16	17	18	19	20
21	22	23	24	25	26	27	28	29	30
31	32	33	34	35	36	37	38	39	40
41	42	43	44	45	46	47	48	49	50
51	52	53	54	55	56	57	58	59	60
61	62	63	64	65	66	67	68	69	70
71	72	73	74	75	76	77	78	79	80
81	82	83	84	85	86	87	88	89	90
91	92	93	94	95	96	97	98	99	100

Multiples of 10

1	2	3	4	5	6	7	8	9	10
11	12	13	14	15	16	17	18	19	20
21	22	23	24	25	26	27	28	29	30
31	32	33	34	35	36	37	38	39	40
41	42	43	44	45	46	47	48	49	50
51	52	53	54	55	56	57	58	59	60
61	62	63	64	65	66	67	68	69	70
71	72	73	74	75	76	77	78	79	80
81	82	83	84	85	86	87	88	89	90
91	92	93	94	95	96	97	98	99	100

Multiples of 12

What happens if you use a different number grid?

1	2	3	4	5	6	7	8	9	10
11	12	13	14	15	16	17	18	19	20
21	22	23	24	25	26	27	28	29	30
31	32	33	34	35	36	37	38	39	40
41	42	43	44	45	46	47	48	49	50
51	52	53	54	55	56	57	58	59	60
61	62	63	64	65	66	67	68	69	70
71	72	73	74	75	76	77	78	79	80
81	82	83	84	85	86	87	88	89	90
91	92	93	94	95	96	97	98	99	100

Try to describe each pattern.

Multiples of 15

www.worldteacherspress.com ©World Teachers Press®

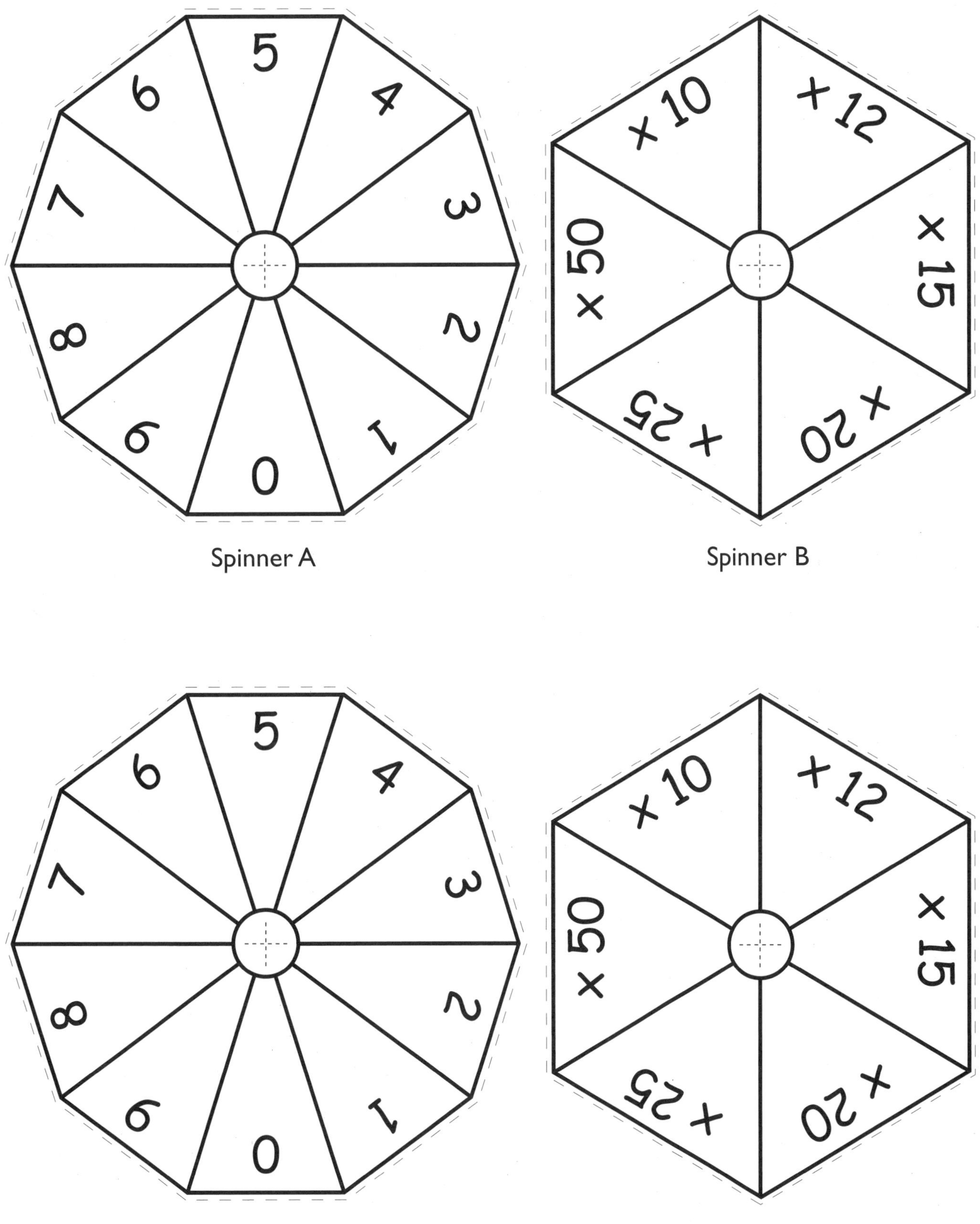

Spinner A

Spinner B

Spinner A

Spinner B

Tables bingo

	x 2	

	x 3	

	x 4	

	x 5	

	x 6	

	x 7	

	x 8	

	x 9	

	x 10	

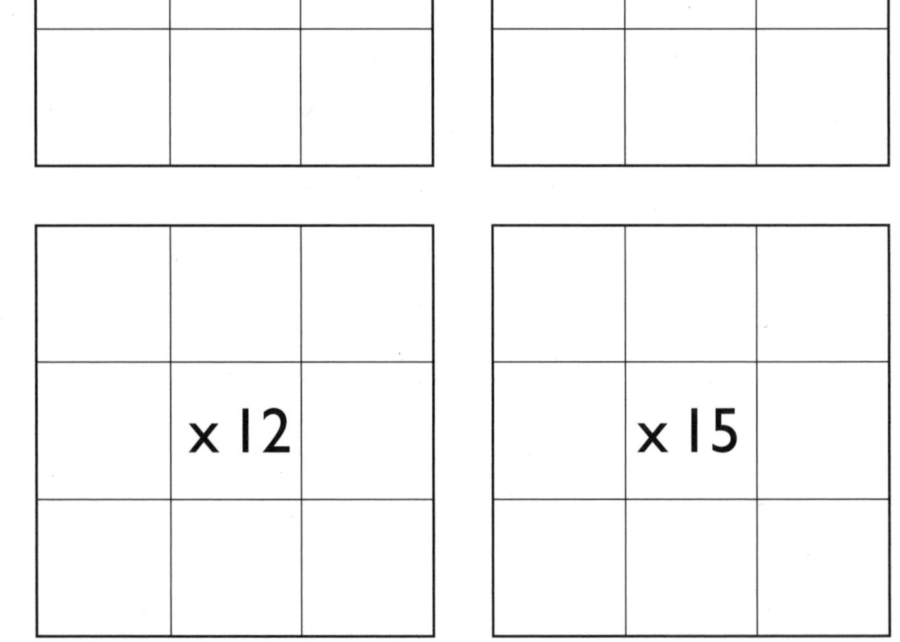

	x 12	

	x 15	

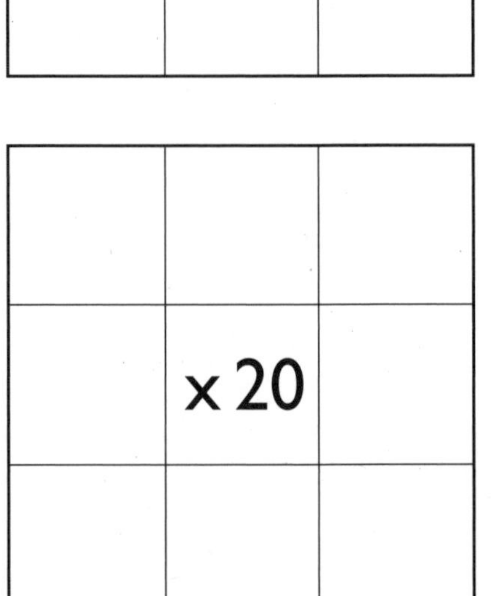

	x 20	

www.worldteacherspress.com ©World Teachers Press®

TABLES

CHECK OF APPROVAL

GREAT EFFORT

in: ______________________________

Name: ______________________________

Date: ______________________________

Signed: ______________________________

Top of the TABLES

Congratulations

TOP!

in

____

Name: ______________________________

Signed: ______________________________

Date: ______________________________

Times Table Terminator

X

Name: ______________________________

Date: ______________________________

Signed: ______________________________

Set A

1. Add 2 to the following:

(a) 14 ____ (c) 20 ____

(b) 4 ____ (d) 18 ____

2. How many groups of 2 in:

(a) 6? ____ (c) 12? ____

(b) 20? ____ (d) 16? ____

3. Subtract 2 from the following:

(a) 18 ____ (c) 12 ____

(b) 8 ____ (d) 10 ____

4. Which multiple of 2 is closest to:

(a) 5? ____ (c) 11? ____

(b) 3? ____ (d) 21? ____

5. Add:

(a) 2 + 2 + 2 + 2 = ____

(b) 2 + 2 + 2 + 2 + 2 = ____

(c) 2 + 2 + 2 + 2 + 2 + 2 + 2 + 2 = ____

6. Complete the number sequence:

0, 2, 4, ____, 8, ____, ____, 14, ____, ____, 20, ____, ____

7. Complete the number sentences:

(a) 2 x ____ = 14

(b) 2 x ____ = 10

(c) 8 x ____ = 16

8. Complete the following:

(a) (2 x 5) + 2 = ____

(b) (2 x 7) – 2 = ____

(c) (2 x 9) + 2 = ____

9. True (✓) False (✗):

(a) 3 x 2 = 6 ☐

(b) 9 x 2 = 16 ☐

(c) 6 x 2 = 14 ☐

(d) 7 x 20 = 140 ☐

(e) 5 x 20 = 100 ☐

(f) 9 x 20 = 160 ☐

(g) 8 ÷ 2 = 4 ☐

(h) 16 ÷ 2 = 8 ☐

(i) 6 ÷ 2 = 3 ☐

(j) 10 ÷ 2 = 6 ☐

10. Say your 2 x tables backwards.

Set B

1. Add 2 to the following:

(a) 10 ____ (c) 18 ____

(b) 8 ____ (d) 20 ____

2. How many groups of 2 in:

(a) 16? ____ (c) 10? ____

(b) 2? ____ (d) 14? ____

3. Subtract 2 from the following:

(a) 4 ____ (c) 20 ____

(b) 14 ____ (d) 22 ____

4. Which multiple of 2 is closest to:

(a) 9? ____ (c) 7? ____

(b) 13? ____ (d) 15? ____

5. Add:

(a) 2 + 2 + 2 + 2 + 2 + 2 + 2 = ____

(b) 2 + 2 + 2 + 2 + 2 + 2 = ____

(c) 2 + 2 + 2 + 2 + 2 + 2 + 2 + 2 + 2 = ____

6. Complete the number sequence:

20, ____, ____, 14, ____, 10, ____, ____, ____, 2, ____

7. Complete the number sentences:

(a) 2 x ____ = 8

(b) 3 x ____ = 6

(c) 2 x ____ = 18

8. Complete the following:

(a) (2 x 6) – 2 = ____

(b) (2 x 8) – 2 = ____

(c) (2 x 2) + 2 = ____

9. True (✓) False (✗):

(a) 8 x 2 = 16 ☐

(b) 5 x 2 = 10 ☐

(c) 7 x 2 = 12 ☐

(d) 4 x 20 = 80 ☐

(e) 3 x 20 = 60 ☐

(f) 6 x 20 = 100 ☐

(g) 8 x 20 = 160 ☐

(h) 18 ÷ 2 = 9 ☐

(i) 14 ÷ 2 = 7 ☐

(j) 12 ÷ 2 = 8 ☐

10. Count by 20s to 200.

www.worldteacherspress.com ©World Teachers Press®

0 x 2 = ____	0 x 20 = ____		
1 x 2 = ____	1 x 20 = ____	6 x 2 = ____	6 x 20 = ____
2 x 2 = ____	2 x 20 = ____	7 x 2 = ____	7 x 20 = ____
3 x 2 = ____	3 x 20 = ____	8 x 2 = ____	8 x 20 = ____
4 x 2 = ____	4 x 20 = ____	9 x 2 = ____	9 x 20 = ____
5 x 2 = ____	5 x 20 = ____	10 x 2 = ____	10 x 20 = ____

1. How many days in:

(a) 2 weeks 4 days? ____________ *(b) 2 weeks 6 days?* ____________

2. Double:

(a) 40 ____ *(b) 100* ____ *(c) 60* ____ *(d) 20* ____

3. Halve:

(a) 160 ____ *(b) 120* ____ *(c) 80* ____ *(d) 180* ____

4. Tiles are 2 cm long. How many can you fit into a space:

(a) 22 cm long? ____ *tiles* *(b) 10 cm long?* ____ *tiles* *(c) 18 cm long?* ____ *tiles*

5. Each bus can carry 20 passengers. How many buses would be needed to carry:

(a) 100 people? ____ *buses* *(b) 180 people?* ____ *buses* *(c) 140 people?* ____ *buses*

6. Complete these:

(a) 17 ÷ 2 = ____ *r* ____ *(b) 13 ÷ 2 =* ____ *r* ____ *(c) 21 ÷ 2 =* ____ *r* ____

7. If you cut these cakes in half, how many slices would you have?

(a) 7 cakes ____ *slices* *(c) 3 cakes* ____ *slices* *(e) 5 cakes* ____ *slices*

(b) 9 cakes ____ *slices* *(d) 4 cakes* ____ *slices* *(f) 2 cakes* ____ *slices*

8. A crate can hold 20 apples. How many apples would fit in:

(a) 6 crates? ____ *apples* *(b) 3 crates?* ____ *apples* *(c) 8 crates?* ____ *apples*

9. Postcards cost 20¢ each. Holly buys 7 and her friend Shari buys 9 cards.

(a) How much money does Holly spend? ____________

(b) How much money does Shari spend? ____________

(c) How much do they spend altogether? ____________

(d) How much change would Holly receive from a $5 bill? ____________

(e) How much change would Shari receive from a $5 bill? ____________

10. Complete the chart.

+ 20	60	180	100	20	140	200	40	160	0	80	120
– 20	100	20	80	200	160	40	180	120	220	60	140

©World Teachers Press® www.worldteacherspress.com

Set A

1. $2 + 6 =$ ____
2. $2 \times 7 =$ ____
3. $12 \div 2 =$ ____
4. $16 - 2 =$ ____
5. $2 + 10 =$ ____
6. $2 \times 0 =$ ____
7. $8 \div 2 =$ ____
8. $20 - 2 =$ ____
9. $2 + 14 =$ ____
10. $2 \times 5 =$ ____
11. $24 \div 2 =$ ____
12. $14 - 2 =$ ____
13. $18 + 2 =$ ____
14. $10 \times 2 =$ ____
15. $10 \div 2 =$ ____
16. $18 - 2 =$ ____
17. $2 + 22 =$ ____
18. $2 \times 4 =$ ____
19. $16 \div 2 =$ ____
20. $22 - 2 =$ ____
21. $4 + 2 =$ ____
22. $8 \times 2 =$ ____
23. $2 \div 2 =$ ____
24. $24 - 2 =$ ____
25. $2 + 8 =$ ____
26. $9 \times 2 =$ ____
27. $18 \div 2 =$ ____
28. $12 - 2 =$ ____
29. $2 + 12 =$ ____
30. $2 \times 6 =$ ____
31. $6 \div 2 =$ ____
32. $8 - 2 =$ ____
33. $16 + 2 =$ ____
34. $0 \times 2 =$ ____
35. $20 \div 2 =$ ____
36. $10 - 2 =$ ____
37. $2 + 20 =$ ____
38. $2 \times 10 =$ ____
39. $14 \div 2 =$ ____
40. $6 - 2 =$ ____

My score: ☐

My time:

____ *mins* ____ *secs*

I'm happy · I'm not happy · OOPS! · I didn't understand

The main area for me to work on is:

____

____

____

____

____

____

____

Set B

1. $2 + 16 =$ ____
2. $2 \times 4 =$ ____
3. $20 \div 2 =$ ____
4. $12 - 2 =$ ____
5. $2 + 20 =$ ____
6. $2 \times 8 =$ ____
7. $10 \div 2 =$ ____
8. $24 - 2 =$ ____
9. $2 + 18 =$ ____
10. $2 \times 10 =$ ____
11. $14 \div 2 =$ ____
12. $6 - 2 =$ ____
13. $22 + 2 =$ ____
14. $2 \times 2 =$ ____
15. $24 \div 2 =$ ____
16. $8 - 2 =$ ____
17. $2 + 14 =$ ____
18. $2 \times 5 =$ ____
19. $6 \div 2 =$ ____
20. $10 - 2 =$ ____
21. $2 + 12 =$ ____
22. $2 \times 9 =$ ____
23. $12 \div 2 =$ ____
24. $14 - 2 =$ ____
25. $2 + 8 =$ ____
26. $7 \times 2 =$ ____
27. $16 \div 2 =$ ____
28. $18 - 2 =$ ____
29. $2 + 10 =$ ____
30. $2 \times 6 =$ ____
31. $2 \div 2 =$ ____
32. $20 - 2 =$ ____
33. $6 + 2 =$ ____
34. $2 \times 20 =$ ____
35. $18 \div 2 =$ ____
36. $22 - 2 =$ ____
37. $2 + 2 =$ ____
38. $0 \times 2 =$ ____
39. $8 \div 2 =$ ____
40. $4 - 2 =$ ____

My score: ☐

My time:

____ *mins* ____ *secs*

I'm happy · I'm not happy · OOPS! · I didn't understand

The main area for me to work on is:

____

____

____

____

____

____

____

www.worldteacherspress.com ©World Teachers Press®

Set A

1. Add 3 to the following:

(a) 12 ______ *(c)* 21 ______

(b) 27 ______ *(d)* 6 ______

2. How many groups of 3 in:

(a) 30? ______ *(c)* 9? ______

(b) 15? ______ *(d)* 18? ______

3. Subtract 3 from the following:

(a) 27 ______ *(c)* 21 ______

(b) 15 ______ *(d)* 12 ______

4. Which multiple of 3 is closest to:

(a) 19? ______ *(c)* 23? ______

(b) 16? ______ *(d)* 31? ______

5. Add:

(a) 3 + 3 + 3 + 3 = ______

(b) 3 + 3 + 3 + 3 + 3 = ______

(c) 3 + 3 + 3 + 3 + 3 + 3 + 3 + 3 = ______

6. Complete the number sequence:

0, 3, 6, ______, 12, ______, ______, 21, ______, ______, 30

7. Complete the number sentences:

(a) 3 x ______ = 27

(b) 6 x ______ = 18

(c) 3 x ______ = 9

8. Complete the following:

(a) (3 x 5) + 3 = ______

(b) (3 x 4) + 3 = ______

(c) (3 x 9) – 3 = ______

9. True (✓) False (✗):

(a) 3 x 3 = 9 ☐

(b) 9 x 3 = 27 ☐

(c) 6 x 3 = 15 ☐

(d) 7 x 30 = 210 ☐

(e) 5 x 30 = 180 ☐

(f) 9 x 30 = 240 ☐

(g) 12 ÷ 3 = 4 ☐

(h) 24 ÷ 3 = 9 ☐

(i) 9 ÷ 3 = 3 ☐

(j) 15 ÷ 5 = 3 ☐

10. Say your 3 x tables backwards.

Set B

1. Add 3 to the following:

(a) 9 ______ *(c)* 3 ______

(b) 18 ______ *(d)* 15 ______

2. How many groups of 3 in:

(a) 21? ______ *(c)* 12? ______

(b) 24? ______ *(d)* 3? ______

3. Subtract 3 from the following:

(a) 9 ______ *(c)* 30 ______

(b) 18 ______ *(d)* 6 ______

4. Which multiple of 3 is closest to:

(a) 8? ______ *(c)* 29? ______

(b) 11? ______ *(d)* 17? ______

5. Add:

(a) 3 + 3 + 3 = ______

(b) 3 + 3 + 3 + 3 + 3 + 3 = ______

(c) 3 + 3 + 3 + 3 + 3 + 3 + 3 + 3 + 3 = ______

6. Complete the number sequence:

30, ______, ______, 21, ______, 15, ______, ______, ______, 3, ______

7. Complete the number sentences:

(a) 3 x ______ = 12

(b) 10 x ______ = 30

(c) 3 x ______ = 15

8. Complete the following:

(a) (3 x 3) + 3 = ______

(b) (3 x 7) + 3 = ______

(c) (3 x 10) – 3 = ______

9. True (✓) False (✗):

(a) 8 x 3 = 21 ☐

(b) 5 x 3 = 15 ☐

(c) 7 x 3 = 21 ☐

(d) 10 x 3 = 30 ☐

(e) 4 x 30 = 120 ☐

(f) 4 x 30 = 90 ☐

(g) 6 x 30 = 180 ☐

(h) 8 x 30 = 240 ☐

(i) 27 ÷ 3 = 9 ☐

(j) 21 ÷ 3 = 6 ☐

10. Count by 30s to 300.

©World Teachers Press® www.worldteacherspress.com

3 x table

0 x 3 = ____	0 x 30 = ____		
1 x 3 = ____	1 x 30 = ____	6 x 3 = ____	6 x 30 = ____
2 x 3 = ____	2 x 30 = ____	7 x 3 = ____	7 x 30 = ____
3 x 3 = ____	3 x 30 = ____	8 x 3 = ____	8 x 30 = ____
4 x 3 = ____	4 x 30 = ____	9 x 3 = ____	9 x 30 = ____
5 x 3 = ____	5 x 30 = ____	10 x 3 = ____	10 x 30 = ____

3

1. How many days in:

(a) 3 weeks 5 days? ____________ *(b) 3 weeks 3 days?* ____________

2. Double:

(a) 60 ____ *(b) 90* ____ *(c) 120* ____ *(d) 15* ____

3. Halve:

(a) 300 ____ *(b) 180* ____ *(c) 240* ____ *(d) 420* ____

4. Tiles are 3 cm long. How many can you fit into a space:

(a) 15 cm long? ____ *tiles* *(b) 30 cm long?* ____ *tiles* *(c) 27 cm long?* ____ *tiles*

5. Each train car can carry 30 passengers. How many cars would be needed to carry:

(a) 150 people? ____ *cars* *(b) 270 people?* ____ *cars* *(c) 120 people?* ____ *cars*

6. Complete these:

(a) $29 \div 3 =$ ____ *r* ____ *(b)* $11 \div 3 =$ ____ *r* ____ *(c)* $19 \div 3 =$ ____ *r* ____

7. If you cut these pizzas into thirds, how many slices would you have?

(a) 4 pizzas ____ *slices* *(c) 9 pizzas* ____ *slices* *(e) 5 pizzas* ____ *slices*

(b) 6 pizzas ____ *slices* *(d) 2 pizzas* ____ *slices* *(f) 10 pizzas* ____ *slices*

8. A box can hold 30 oranges. How many oranges would fit in:

(a) 6 boxes? ____ *oranges* *(b) 4 boxes?* ____ *oranges* *(c) 7 boxes?* ____ *oranges*

9. Eggs cost 30¢ each. Josh buys 8 and Lauren buys 5 eggs.

(a) How much money does Josh spend? ____________

(b) How much money does Lauren spend? ____________

(c) How much more does Josh spend than Lauren? ____________

(d) How much change would Josh get if he paid with a $5 bill? ____________

(e) How much change would Lauren get if she paid with a $5 bill? ____________

10. Complete the chart.

+ 30	120	30	270	180	90	300	150	210	60	240	0
– 30	190	300	210	120	60	270	90	240	330	30	150

www.worldteacherspress.com ©World Teachers Press®

Set A

1. 3 + 6 = ____	21. 3 + 3 = ____
2. 3 x 7 = ____	22. 5 x 3 = ____
3. 30 ÷ 3 = ____	23. 27 ÷ 3 = ____
4. 18 – 3 = ____	24. 21 – 3 = ____
5. 3 + 9 = ____	25. 3 + 21 = ____
6. 3 x 8 = ____	26. 4 x 3 = ____
7. 12 ÷ 3 = ____	27. 15 ÷ 3 = ____
8. 27 – 3 = ____	28. 15 – 3 = ____
9. 3 + 12 = ____	29. 27 + 3 = ____
10. 3 x 9 = ____	30. 3 x 2 = ____
11. 21 ÷ 3 = ____	31. 6 ÷ 3 = ____
12. 24 – 3 = ____	32. 9 – 3 = ____
13. 3 + 15 = ____	33. 24 + 3 = ____
14. 3 x 0 = ____	34. 3 x 3 = ____
15. 18 ÷ 3 = ____	35. 9 ÷ 3 = ____
16. 30 – 3 = ____	36. 3 – 3 = ____
17. 3 + 18 = ____	37. 30 + 3 = ____
18. 3 x 6 = ____	38. 3 x 0 = ____
19. 24 ÷ 3 = ____	39. 3 ÷ 3 = ____
20. 12 – 3 = ____	40. 6 – 3 = ____

My score: ☐

My time:

____ mins ____ secs

I'm happy · I'm not happy · OOPS! · I didn't understand

The main area for me to work on is:

____

____

____

____

____

____

____

Set B

1. 3 + 9 = ____	21. 3 + 18 = ____
2. 3 x 6 = ____	22. 3 x 7 = ____
3. 15 ÷ 3 = ____	23. 21 ÷ 3 = ____
4. 27 – 3 = ____	24. 33 – 3 = ____
5. 3 + 21 = ____	25. 3 + 24 = ____
6. 3 x 0 = ____	26. 10 x 3 = ____
7. 24 ÷ 3 = ____	27. 6 ÷ 3 = ____
8. 18 – 3 = ____	28. 21 – 3 = ____
9. 3 + 27 = ____	29. 3 + 30 = ____
10. 3 x 8 = ____	30. 3 x 30 = ____
11. 9 ÷ 3 = ____	31. 18 ÷ 3 = ____
12. 30 – 3 = ____	32. 24 – 3 = ____
13. 3 + 6 = ____	33. 3 + 3 = ____
14. 3 x 2 = ____	34. 9 x 3 = ____
15. 30 ÷ 10= ____	35. 12 ÷ 3 = ____
16. 9 – 3 = ____	36. 36 – 3 = ____
17. 12 + 3 = ____	37. 3 + 15 = ____
18. 3 x 4 = ____	38. 3 x 3 = ____
19. 3 ÷ 3 = ____	39. 27 ÷ 3 = ____
20. 12 – 3 = ____	40. 15 – 3 = ____

My score: ☐

My time:

____ mins ____ secs

I'm happy · I'm not happy · OOPS! · I didn't understand

The main area for me to work on is:

____

____

____

____

____

____

____

©World Teachers Press® www.worldteacherspress.com

Set A

1. Add 4 to the following:

(a) 16 ______ (c) 32 ______

(b) 8 ______ (d) 28 ______

2. How many groups of 4 in:

(a) 24? ______ (c) 16? ______

(b) 32? ______ (d) 8? ______

3. Subtract 4 from the following:

(a) 28 ______ (c) 12 ______

(b) 36 ______ (d) 20 ______

4. Which multiple of 4 is closest to:

(a) 39? ______ (c) 23? ______

(b) 17? ______ (d) 31? ______

5. Add:

(a) 4 + 4 + 4 = ______

(b) 4 + 4 + 4 + 4 + 4 = ______

(c) 4 + 4 + 4 + 4 + 4 + 4 + 4 + 4 = ______

6. Complete the number sequence:

0, 4, ______, 12, ______, 20, ______, ______, 32, ______, 40

7. Complete the number sentences:

(a) 4 x ______ = 20

(b) 4 x ______ = 0

(c) 4 x ______ = 12

8. Complete the following:

(a) (4 x 7) + 4 = ______

(b) (4 x 5) + 4 = ______

(c) (4 x 10) – 4 = ______

9. True (✓) False (✗):

(a) 3 x 4 = 12 ☐

(b) 9 x 4 = 32 ☐

(c) 6 x 4 = 24 ☐

(d) 7 x 40 = 320 ☐

(e) 5 x 40 = 200 ☐

(f) 9 x 40 = 360 ☐

(g) 16 ÷ 4 = 5 ☐

(h) 32 ÷ 4 = 8 ☐

(i) 12 ÷ 4 = 3 ☐

(j) 20 ÷ 4 = 4 ☐

10. Say your 4 x tables backwards.

Set B

1. Add 4 to the following:

(a) 20 ______ (c) 36 ______

(b) 24 ______ (d) 40 ______

2. How many groups of 4 in:

(a) 40? ______ c) 12? ______

(b) 20? ______ (d) 36? ______

3. Subtract 4 from the following:

(a) 16 ______ (c) 8 ______

(b) 32 ______ (d) 40 ______

4. Which multiple of 4 is closest to:

(a) 11? ______ (c) 19? ______

(b) 3? ______ (d) 27? ______

5. Add:

(a) 4 + 4 + 4 + 4 + 4 + 4 + 4 = ______

(b) 4 + 4 + 4 + 4 + 4 + 4 = ______

(c) 4 + 4 + 4 + 4 + 4 + 4 + 4 + 4 + 4 + 4 = ______

6. Complete the number sequence:

40, ______, ______, 28, ______, ______, 16, ______, ______, 4, ______

7. Complete the number sentences:

(a) 4 x ______ = 36

(b) 8 x ______ = 32

(c) 4 x ______ = 28

8. Complete the following:

(a) (4 x 4) + 4 = ______

(b) (4 x 8) + 4 = ______

(c) (4 x 6) – 4 = ______

9. True (✓) False (✗):

(a) 8 x 4 = 36 ☐

(b) 5 x 4 = 24 ☐

(c) 7 x 4 = 28 ☐

(d) 10 x 4 = 40 ☐

(e) 3 x 40 = 120 ☐

(f) 6 x 40 = 240 ☐

(g) 8 x 40 = 280 ☐

(h) 36 ÷ 4 = 8 ☐

(i) 28 ÷ 4 = 7 ☐

(j) 24 ÷ 4 = 6 ☐

10. Count by 40s to 400.

www.worldteacherspress.com ©World Teachers Press®

4 x table

0 x 4 = ____	0 x 40 = ____		
1 x 4 = ____	1 x 40 = ____	6 x 4 = ____	6 x 40 = ____
2 x 4 = ____	2 x 40 = ____	7 x 4 = ____	7 x 40 = ____
3 x 4 = ____	3 x 40 = ____	8 x 4 = ____	8 x 40 = ____
4 x 4 = ____	4 x 40 = ____	9 x 4 = ____	9 x 40 = ____
5 x 4 = ____	5 x 40 = ____	10 x 4 = ____	10 x 40 = ____

1. How many days in:

(a) 4 weeks 3 days? ____________ *(b) 4 weeks 6 days?* ____________

2. Double:

(a) 80 ____ *(b) 120* ____ *(c) 40* ____ *(d) 160* ____

3. Halve:

(a) 400 ____ *(b) 320* ____ *(c) 80* ____ *(d) 280* ____

4. Tiles are 4 cm long. How many can you fit into a space:

(a) 20 cm long? ____ *tiles* *(b) 40 cm long?* ____ *tiles* *(c) 36 cm long?* ____ *tiles*

5. Each train car can carry 40 passengers. How many cars would be needed to carry:

(a) 160 people? ____ *cars* *(b) 360 people?* ____ *cars* *(c) 280 people?* ____ *cars*

6. Complete these:

(a) 25 ÷ 4 = ____ *r* ____ *(b) 30 ÷ 4 =* ____ *r* ____ *(c) 18 ÷ 4 =* ____ *r* ____

7. If you cut these pies into fourths, how many slices would you have?

(a) 3 pies ____ *slices* *(c) 7 pies* ____ *slices* *(e) 5 pies* ____ *slices*

(b) 9 pies ____ *slices* *(d) 4 pies* ____ *slices* *(f) 8 pies* ____ *slices*

8. A crate can hold 40 mangoes. How many mangoes would fit in:

(a) 5 crates? ____ *mangoes* *(b) 3 crates?* ____ *mangoes* *(c) 8 crates?* ____ *mangoes*

9. Collectable cards cost 40¢ each. Kellan buys 9 and his friend Nathan buys 6 cards.

(a) How much money does Kellan spend? ____________

(b) How much money does Nathan spend? ____________

(c) How much do they spend altogether? ____________

(d) How much change would Kellan receive from a $10 bill? ____________

(e) How much change would Nathan receive from a $10 bill? ____________

10. Complete the chart.

+ 40	240	80	160	280	360	40	120	200	320	400	0
– 40	80	320	200	40	360	280	120	400	240	160	440

©World Teachers Press® www.worldteacherspress.com

Set A

1. 4 + 4 = ______
2. 4 x 5 = ______
3. 40 ÷ 4 = ______
4. 12 – 4 = ______
5. 4 + 12 = ______
6. 4 x 7 = ______
7. 16 ÷ 4 = ______
8. 24 – 4 = ______
9. 4 + 28 = ______
10. 4 x 3 = ______
11. 8 ÷ 4 = ______
12. 8 – 4 = ______
13. 4 + 8 = ______
14. 4 x 9 = ______
15. 12 ÷ 4 = ______
16. 36 – 4 = ______
17. 4 + 20 = ______
18. 4 x 10 = ______
19. 32 ÷ 4 = ______
20. 40 – 4 = ______
21. 4 + 16 = ______
22. 4 x 8 = ______
23. 20 ÷ 4 = ______
24. 28 – 4 = ______
25. 32 + 4 = ______
26. 4 x 4 = ______
27. 28 ÷ 4 = ______
28. 16 – 4 = ______
29. 40 + 4 = ______
30. 4 x 6 = ______
31. 36 ÷ 4 = ______
32. 20 – 4 = ______
33. 24 + 4 = ______
34. 2 x 4 = ______
35. 24 ÷ 4 = ______
36. 4 – 4 = ______
37. 36 + 4 = ______
38. 4 x 0 = ______
39. 40 ÷ 4 = ______
40. 32 – 4 = ______

My score:

My time:

______ mins ______ secs

I'm happy

I'm not happy

OOPS!

I didn't understand

The main area for me to work on is:

____

____

____

____

____

____

____

Set B

1. 4 + 12 = ______
2. 4 x 7 = ______
3. 24 ÷ 4 = ______
4. 8 – 4 = ______
5. 4 + 20 = ______
6. 4 x 9 = ______
7. 8 ÷ 4 = ______
8. 32 – 4 = ______
9. 4 + 16 = ______
10. 4 x 3 = ______
11. 32 ÷ 4 = ______
12. 44 – 4 = ______
13. 4 + 8 = ______
14. 4 x 6 = ______
15. 12 ÷ 4 = ______
16. 36 – 4 = ______
17. 4 + 4 = ______
18. 4 x 10 = ______
19. 16 ÷ 4 = ______
20. 24 – 4 = ______
21. 32 + 4 = ______
22. 8 x 40 = ______
23. 28 ÷ 4 = ______
24. 16 – 4 = ______
25. 4 + 40 = ______
26. 4 x 8 = ______
27. 20 ÷ 4 = ______
28. 20 – 4 = ______
29. 40 + 4 = ______
30. 5 x 40 = ______
31. 36 ÷ 4 = ______
32. 12 – 4 = ______
33. 24 + 4 = ______
34. 4 x 0 = ______
35. 4 ÷ 4 = ______
36. 28 – 4 = ______
37. 4 + 36 = ______
38. 4 x 4 = ______
39. 40 ÷ 4 = ______
40. 40 – 4 = ______

My score:

My time:

______ mins ______ secs

I'm happy

I'm not happy

OOPS!

I didn't understand

The main area for me to work on is:

____

____

____

____

____

____

____

www.worldteacherspress.com ©World Teachers Press®

Set A

1. Add 5 to the following:

(a) 15 ______ *(c) 35* ______

(b) 20 ______ *(d) 30* ______

2. How many groups of 5 in:

(a) 10? ______ *(c) 45?* ______

(b) 25? ______ *(d) 30?* ______

3. Subtract 5 from the following:

(a) 30 ______ *(c) 50* ______

(b) 25 ______ *(d) 15* ______

4. Which multiple of 5 is closest to:

(a) 47? ______ *(c) 14?* ______

(b) 33? ______ *(d) 39?* ______

5. Add:

(a) $5 + 5 + 5 =$ ______

(b) $5 + 5 + 5 + 5 + 5 + 5 =$ ______

(c) $5 + 5 + 5 + 5 + 5 + 5 + 5 + 5 =$ ______

6. Complete the number sequence:

0, 5, ______, *15,* ______, ______, *30,* ______, *40,* ______, ______

7. Complete the number sentences:

(a) 5 x ______ = 50

(b) 5 x ______ = 0

(c) 5 x ______ = 30

8. Complete the following:

(a) (5 x 6) + 5 = ______

(b) (5 x 3) + 5 = ______

(c) (5 x 8) – 5 = ______

9. True (✓) False (✗):

(a) 3 x 5 = 15 ☐

(b) 9 x 5 = 45 ☐

(c) 6 x 5 = 35 ☐

(d) 8 x 5 = 40 ☐

(e) 9 x 50 = 400 ☐

(f) 4 x 50 = 200 ☐

(g) 3 x 50 = 150 ☐

(h) 20 ÷ 5 = 4 ☐

(i) 40 ÷ 5 = 10 ☐

(j) 15 ÷ 5 = 3 ☐

10. Say your 5 x tables backwards.

Set B

1. Add 5 to the following:

(a) 5 ______ *(c) 40* ______

(b) 25 ______ *(d) 10* ______

2. How many groups of 5 in:

(a) 50? ______ *(c) 20?* ______

(b) 35? ______ *(d) 40?* ______

3. Subtract 5 from the following:

(a) 10 ______ *(c) 20* ______

(b) 45 ______ *(d) 35* ______

4. Which multiple of 5 is closest to:

(a) 52? ______ *(c) 8?* ______

(b) 27? ______ *(d) 17?* ______

5. Add:

(a) $5 + 5 + 5 + 5 + 5 + 5 + 5 =$ ______

(b) $5 + 5 + 5 + 5 =$ ______

(c) $5 + 5 + 5 + 5 + 5 + 5 + 5 + 5 + 5 + 5 =$ ______

6. Complete the number sequence:

______, *45,* ______, *35,* ______, ______, *20, 15,* ______, ______,

7. Complete the number sentences:

(a) 5 x ______ = 45

(b) 5 x ______ = 10

(c) 8 x ______ = 40

8. Complete the following:

(a) (5 x 9) + 5 = ______

(b) (5 x 10) – 5 = ______

(c) (5 x 7) – 5 = ______

9. True (✓) False (✗):

(a) 5 x 5 = 20 ☐

(b) 7 x 5 = 30 ☐

(c) 10 x 5 = 50 ☐

(d) 7 x 50 = 300 ☐

(e) 5 x 50 = 250 ☐

(f) 6 x 50 = 350 ☐

(g) 8 x 50 = 400 ☐

(h) 25 ÷ 5 = 5 ☐

(i) 45 ÷ 5 = 8 ☐

(j) 30 ÷ 5 = 6 ☐

10. Count by 50s to 500.

5 x table

0 x 5 = ____	0 x 50 = ____		
1 x 5 = ____	1 x 50 = ____	6 x 5 = ____	6 x 50 = ____
2 x 5 = ____	2 x 50 = ____	7 x 5 = ____	7 x 50 = ____
3 x 5 = ____	3 x 50 = ____	8 x 5 = ____	8 x 50 = ____
4 x 5 = ____	4 x 50 = ____	9 x 5 = ____	9 x 50 = ____
5 x 5 = ____	5 x 50 = ____	10 x 5 = ____	10 x 50 = ____

1. How many days in:

(a) 5 weeks 5 days? ____________ *(b) 5 weeks 2 days?* ____________

2. Double:

(a) 150 ____ *(b) 250* ____ *(c) 25* ____ *(d) 100* ____

3. Halve:

(a) 200 ____ *(b) 300* ____ *(c) 100* ____ *(d) 500* ____

4. Candies cost 5¢ each. How many can you buy with:

(a) 30¢? ____ *(b) 45¢?* ____ *(c) 50¢?* ____ *(d) 25¢?* ____

5. Each plane can carry 50 passengers. How many planes would be needed to carry:

(a) 350 people? ____ *planes* *(b) 200 people?* ____ *planes* *(c) 150 people?* ____ *planes*

6. Complete these:

(a) 27 ÷ 5 = ____ *r* ____ *(b) 39 ÷ 5 =* ____ *r* ____ *(c) 48 ÷ 5 =* ____ *r* ____

7. If you cut these oranges into fifths, how many slices would you have?

(a) 6 oranges ____ *slices* *(c) 4 oranges* ____ *slices* *(e) 2 oranges* ____ *slices*

(b) 5 oranges ____ *slices* *(d) 10 oranges* ____ *slices* *(f) 20 oranges* ____ *slices*

8. Each cattle truck can hold 50 cattle. How many cattle would fit in:

(a) 3 trucks? ____ *cattle* *(b) 8 trucks?* ____ *cattle* *(c) 6 trucks?* ____ *cattle*

9. Phone calls cost 50¢ each. Troy made 7 calls and Asha made 9 calls.

(a) How much money does Troy spend? ____________

(b) How much money does Asha spend? ____________

(c) How much more does Asha spend than Troy? ____________

(d) How much change would Troy receive from a $5 bill? ____________

(e) How much change would Asha receive from a $5 bill? ____________

10. Complete the chart.

+ 50	200	50	450	300	150	500	250	350	100	400	0
– 50	300	450	50	100	350	250	500	150	400	200	50

www.worldteacherspress.com ©World Teachers Press®

Set A

1. 5 + 5 = ______
2. 5 x 8 = ______
3. 15 ÷ 5 = ______
4. 30 – 5 = ______
5. 5 + 40 = ______
6. 5 x 6 = ______
7. 35 ÷ 5 = ______
8. 60 – 5 = ______
9. 5 + 20 = ______
10. 5 x 2 = ______
11. 20 ÷ 5 = ______
12. 40 – 5 = ______
13. 25 + 5 = ______
14. 9 x 5 = ______
15. 50 ÷ 5 = ______
16. 10 – 5 = ______
17. 5 + 50 = ______
18. 5 x 3 = ______
19. 10 ÷ 5 = ______
20. 25 – 5 = ______
21. 5 + 35 = ______
22. 5 x 10 = ______
23. 25 ÷ 5 = ______
24. 45 – 5 = ______
25. 30 + 5 = ______
26. 5 x 4 = ______
27. 30 ÷ 5 = ______
28. 55 – 5 = ______
29. 5 + 10 = ______
30. 5 x 7 = ______
31. 45 ÷ 5 = ______
32. 20 – 5 = ______
33. 5 + 45 = ______
34. 5 x 5 = ______
35. 5 ÷ 5 = ______
36. 15 – 5 = ______
37. 5 – 5 = ______
38. 3 x 50 = ______
39. 40 ÷ 5 = ______
40. 35 – 5 = ______

My score: ☐

My time:

_____ mins _____ secs

I'm happy · I'm not happy · OOPS! · I didn't understand

The main area for me to work on is:

____

____

____

____

____

____

____

Set B

1. 5 + 45 = ______
2. 5 x 5 = ______
3. 45 ÷ 5 = ______
4. 15 – 5 = ______
5. 5 + 5 = ______
6. 5 x 0 = ______
7. 5 ÷ 5 = ______
8. 20 – 5 = ______
9. 5 + 10 = ______
10. 5 x 7 = ______
11. 40 ÷ 5 = ______
12. 35 – 5 = ______
13. 5 + 30 = ______
14. 5 x 3 = ______
15. 25 ÷ 5 = ______
16. 55 – 5 = ______
17. 5 + 35 = ______
18. 5 x 10 = ______
19. 10 ÷ 5 = ______
20. 45 – 5 = ______
21. 5 + 50 = ______
22. 5 x 3 = ______
23. 50 ÷ 5 = ______
24. 25 – 5 = ______
25. 5 + 25 = ______
26. 5 x 9 = ______
27. 20 ÷ 5 = ______
28. 10 – 5 = ______
29. 5 + 20 = ______
30. 5 x 2 = ______
31. 35 ÷ 5 = ______
32. 40 – 5 = ______
33. 5 + 45 = ______
34. 6 x 5 = ______
35. 15 ÷ 5 = ______
36. 60 – 5 = ______
37. 15 + 5 = ______
38. 5 x 8 = ______
39. 40 ÷ 5 = ______
40. 30 – 5 = ______

My score: ☐

My time:

_____ mins _____ secs

I'm happy · I'm not happy · OOPS! · I didn't understand

The main area for me to work on is:

____

____

____

____

____

____

____

©World Teachers Press® www.worldteacherspress.com

Set A

1. Add 6 to the following:

(a) 36 ______ (c) 48 ______
(b) 24 ______ (d) 18 ______

2. How many groups of 6 in:

(a) 18? ______ (c) 30? ______
(b) 54? ______ (d) 60? ______

3. Subtract 6 from the following:

(a) 12 ______ (c) 54 ______
(b) 30 ______ (d) 24 ______

4. Which multiple of 6 is closest to:

(a) 58? ______ (c) 34? ______
(b) 19? ______ (d) 53? ______

5. Add:

(a) 6 + 6 + 6 + 6 + 6 + 6 + 6 + 6 = ______
(b) 6 + 6 + 6 + 6 + 6 = ______
(c) 6 + 6 + 6 + 6 + 6 + 6 = ______

6. Complete the number sequence:

0, 6, ______, ______, 24, ______, ______, 42, ______, ______, ______

7. Complete the number sentences:

(a) 6 x ______ = 54
(b) 3 x ______ = 18
(c) 6 x ______ = 24

8. Complete the following:

(a) (6 x 7) + 6 = ______
(b) (6 x 3) + 6 = ______
(c) (6 x 10) – 6 = ______

9. True (✓) False (✗):

(a) 3 x 6 = 21 ☐
(b) 9 x 6 = 54 ☐
(c) 6 x 6 = 30 ☐
(d) 8 x 6 = 48 ☐
(e) 7 x 60 = 420 ☐
(f) 5 x 60 = 300 ☐
(g) 9 x 60 = 480 ☐
(h) 54 ÷ 6 = 9 ☐
(i) 42 ÷ 6 = 7 ☐
(j) 36 ÷ 6 = 5 ☐

10. Say your 6 x tables backwards.

Set B

1. Add 6 to the following:

(a) 12 ______ (c) 42 ______
(b) 54 ______ (d) 30 ______

2. How many groups of 6 in:

(a) 36? ______ (c) 6? ______
(b) 42? ______ (d) 24? ______

3. Subtract 6 from the following:

(a) 18 ______ (c) 60 ______
(b) 48 ______ (d) 42 ______

4. Which multiple of 6 is closest to:

(a) 44? ______ (c) 25? ______
(b) 46? ______ (d) 14? ______

5. Add:

(a) 6 + 6 + 6 = ______
(b) 6 + 6 + 6 + 6 + 6 + 6 + 6 = ______
(c) 6 + 6 + 6 + 6 + 6 + 6 + 6 + 6 + 6 = ______

6. Complete the number sequence:

60, ______, ______, 42, ______, ______, 24, 18, ______, ______, 0

7. Complete the number sentences:

(a) 6 x ______ = 36
(b) 6 x ______ = 0
(c) 6 x ______ = 12

8. Complete the following:

(a) (6 x 9) + 6 = ______
(b) (6 x 4) – 6 = ______
(c) (6 x 6) – 6 = ______

9. True (✓) False (✗):

(a) 5 x 6 = 30 ☐
(b) 7 x 6 = 36 ☐
(c) 10 x 6 = 60 ☐
(d) 4 x 60 = 240 ☐
(e) 3 x 60 = 120 ☐
(f) 6 x 60 = 360 ☐
(g) 54 ÷ 6 = 4 ☐
(h) 48 ÷ 6 = 8 ☐
(i) 18 ÷ 6 = 3 ☐
(j) 30 ÷ 6 = 6 ☐

10. Count by 60s to 600.

www.worldteacherspress.com ©World Teachers Press®

0 x 6 = ____	0 x 60 = ____		
1 x 6 = ____	1 x 60 = ____	6 x 6 = ____	6 x 60 = ____
2 x 6 = ____	2 x 60 = ____	7 x 6 = ____	7 x 60 = ____
3 x 6 = ____	3 x 60 = ____	8 x 6 = ____	8 x 60 = ____
4 x 6 = ____	4 x 60 = ____	9 x 6 = ____	9 x 60 = ____
5 x 6 = ____	5 x 60 = ____	10 x 6 = ____	10 x 60 = ____

1. How many days in:

(a) 6 weeks 4 days? ________ *(b) 6 weeks 3 days?* ________

2. Double:

(a) 12 ____ *(b) 180* ____ *(c) 30* ____ *(d) 300* ____

3. Halve:

(a) 120 ____ *(b) 240* ____ *(c) 480* ____ *(d) 60* ____

4. Tiles are 6 cm long. How many can you fit into a space:

(a) 42 cm long? ____ *tiles* *(b) 54 cm long?* ____ *tiles* *(c) 60 cm long?* ____ *tiles*

5. Each bus can carry 60 passengers. How many buses would be needed to carry:

(a) 300 people? ____ *buses* *(b) 480 people?* ____ *buses* *(c) 180 people?* ____ *buses*

6. Complete these:

(a) $26 \div 6 =$ ____ *r* ____ *(b)* $57 \div 6 =$ ____ *r* ____ *(c)* $40 \div 6 =$ ____ *r* ____

7. If you cut these cakes into sixths, how many slices would you have?

(a) 3 cakes ____ *slices* *(c) 5 cakes* ____ *slices* *(e) 6 cakes* ____ *slices*

(b) 8 cakes ____ *slices* *(d) 10 cakes* ____ *slices* *(f) 9 cakes* ____ *slices*

8. A crate can hold 60 apples. How many apples would fit in:

(a) 2 crates? ____ *apples* *(b) 6 crates?* ____ *apples* *(c) 9 crates?* ____ *apples*

9. Cards cost 60¢ each. Ben buys 6 cards and his friend Josh buys 9.

(a) How much money does Ben spend? ________

(b) How much money does Josh spend? ________

(c) How much do Ben and Josh spend altogether? ________

(d) How much change would Ben receive from a $10 bill? ________

(e) How much change would Josh receive from a $10 bill? ________

10. Chutes 'n' Ladders

																			Finish
81	82	83	84	85	86	87	88	89	90	91	92	93	94	95	96	97	98	99	100
80	79	78	77	76	75	74	73	72	71	70	69	68	67	66	65	64	63	62	61
41	42	43	44	45	46	47	48	49	50	51	52	53	54	55	56	57	58	59	60
40	39	38	37	36	35	34	33	32	31	30	29	28	27	26	25	24	23	22	21
Start 1	2	3	4	5	6	7	8	9	10	11	12	13	14	15	16	17	18	19	20

©World Teachers Press® www.worldteacherspress.com

6 x table

Set A

1. 6 + 6 = ____
2. 6 x 7 = ____
3. 24 ÷ 6 = ____
4. 30 – 6 = ____
5. 6 + 36 = ____
6. 6 x 10 = ____
7. 48 ÷ 6 = ____
8. 18 – 6 = ____
9. 12 + 6 = ____
10. 6 x 4 = ____
11. 12 ÷ 6 = ____
12. 36 – 6 = ____
13. 6 + 54 = ____
14. 6 x 8 = ____
15. 54 ÷ 6 = ____
16. 24 – 6 = ____
17. 6 + 30 = ____
18. 6 x 0 = ____
19. 60 ÷ 6 = ____
20. 48 – 6 = ____
21. 60 + 6 = ____
22. 6 x 9 = ____
23. 18 ÷ 6 = ____
24. 12 – 6 = ____
25. 6 + 18 = ____
26. 6 x 60 = ____
27. 6 ÷ 6 = ____
28. 42 – 6 = ____
29. 6 + 42 = ____
30. 6 x 5 = ____
31. 42 ÷ 6 = ____
32. 54 – 6 = ____
33. 6 + 48 = ____
34. 6 x 3 = ____
35. 30 ÷ 6 = ____
36. 60 – 6 = ____
37. 24 + 6 = ____
38. 6 x 6 = ____
39. 36 ÷ 6 = ____
40. 66 – 6 = ____

My score:

My time:

____ mins ____ secs

I'm happy · I'm not happy · OOPS! · I didn't understand

The main area for me to work on is:

____

____

____

____

____

____

____

Set B

1. 54 + 6 = ____
2. 6 x 3 = ____
3. 18 ÷ 6 = ____
4. 54 – 6 = ____
5. 6 + 30 = ____
6. 6 x 6 = ____
7. 6 ÷ 6 = ____
8. 66 – 6 = ____
9. 6 + 0 = ____
10. 6 x 5 = ____
11. 42 ÷ 6 = ____
12. 60 – 6 = ____
13. 6 + 18 = ____
14. 6 x 0 = ____
15. 36 ÷ 6 = ____
16. 42 – 6 = ____
17. 6 + 42 = ____
18. 6 x 9 = ____
19. 30 ÷ 6 = ____
20. 12 – 6 = ____
21. 6 + 48 = ____
22. 3 x 60 = ____
23. 12 ÷ 6 = ____
24. 48 – 6 = ____
25. 6 + 24 = ____
26. 6 x 8 = ____
27. 60 ÷ 6 = ____
28. 24 – 6 = ____
29. 6 + 12 = ____
30. 6 x 4 = ____
31. 54 ÷ 6 = ____
32. 36 – 6 = ____
33. 6 + 36 = ____
34. 6 x 10 = ____
35. 24 ÷ 6 = ____
36. 18 – 6 = ____
37. 6 + 6 = ____
38. 6 x 7 = ____
39. 48 ÷ 6 = ____
40. 30 – 6 = ____

My score:

My time:

____ mins ____ secs

I'm happy · I'm not happy · OOPS! · I didn't understand

The main area for me to work on is:

____

____

____

____

____

____

____

www.worldteacherspress.com ©World Teachers Press®

Set A

1. Add 7 to the following:

(a) 21 ____ (b) 63 ____ (c) 56 ____ (d) 7 ____

2. How many groups of 7 in:

(a) 14? ____ (b) 56? ____ (c) 35? ____ (d) 42? ____

3. Subtract 7 from the following:

(a) 70 ____ (b) 35 ____ (c) 14 ____ (d) 42 ____

4. Which multiple of 7 is closest to:

(a) 22? ____ (b) 17? ____ (c) 40? ____ (d) 9? ____

5. Add:

(a) 7 + 7 + 7 = ____

(b) 7 + 7 + 7 + 7 + 7 + 7 + 7 = ____

(c) 7 + 7 + 7 + 7 + 7 = ____

6. Complete the number sequence:

0, 7, ____, ____, ____, 35, 42, ____, ____, 63, ____

7. Complete the number sentences:

(a) 7 x ____ = 56

(b) 2 x ____ = 14

(c) 7 x ____ = 70

8. Complete the following:

(a) (7 x 9) + 7 = ____

(b) (7 x 3) + 7 = ____

(c) (7 x 4) – 7 = ____

9. True (✓) False (✗):

(a) 3 x 7 = 21 ☐

(b) 9 x 7 = 56 ☐

(c) 6 x 7 = 49 ☐

(d) 4 x 70 = 280 ☐

(e) 3 x 70 = 210 ☐

(f) 6 x 70 = 420 ☐

(g) 8 x 70 = 630 ☐

(h) 42 ÷ 7 = 6 ☐

(i) 49 ÷ 7 = 8 ☐

(j) 63 ÷ 7 = 9 ☐

10. Say your 7 x tables backwards.

Set B

1. Add 7 to the following:

(a) 14 ____ (b) 28 ____ (c) 42 ____ (d) 63 ____

2. How many groups of 7 in:

(a) 7? ____ (b) 63? ____ (c) 70? ____ (d) 21? ____

3. Subtract 7 from the following:

(a) 49 ____ (b) 21 ____ (c) 63 ____ (d) 56 ____

4. Which multiple of 7 is closest to:

(a) 30? ____ (b) 12? ____ (c) 59? ____ (d) 51? ____

5. Add:

(a) 7 + 7 + 7 + 7 + 7 + 7 + 7 + 7 + 7 = ____

(b) 7 + 7 + 7 + 7 = ____

(c) 7 + 7 + 7 + 7 + 7 + 7 + 7 + 7 = ____

6. Complete the number sequence:

____, 63, ____, 49, ____, ____, 28, ____, ____, 7, 0

7. Complete the number sentences:

(a) 7 x ____ = 28

(b) 9 x ____ = 63

(c) 7 x ____ = 42

8. Complete the following:

(a) (7 x 8) + 7 = ____

(b) (7 x 10) – 7 = ____

(c) (7 x 5) – 7 = ____

9. True (✓) False (✗):

(a) 8 x 7 = 56 ☐

(b) 5 x 7 = 42 ☐

(c) 7 x 7 = 49 ☐

(d) 10 x 7 = 70 ☐

(e) 7 x 70 = 560 ☐

(f) 5 x 70 = 350 ☐

(g) 9 x 70 = 560 ☐

(h) 28 ÷ 7 = 4 ☐

(i) 56 ÷ 7 = 8 ☐

(j) 21 ÷ 7 = 3 ☐

10. Count by 70s to 700.

©World Teachers Press® www.worldteacherspress.com

0 x 7 = ______	0 x 70 = ______		
1 x 7 = ______	1 x 70 = ______	6 x 7 = ______	6 x 70 = ______
2 x 7 = ______	2 x 70 = ______	7 x 7 = ______	7 x 70 = ______
3 x 7 = ______	3 x 70 = ______	8 x 7 = ______	8 x 70 = ______
4 x 7 = ______	4 x 70 = ______	9 x 7 = ______	9 x 70 = ______
5 x 7 = ______	5 x 70 = ______	10 x 7 = ______	10 x 70 = ______

1. How many days in:

(a) 7 weeks 5 days? ______________ *(b) 7 weeks 4 days?* ______________

2. Double:

(a) 210 ______ *(b) 280* ______ *(c) 35* ______ *(d) 21* ______

3. Halve:

(a) 700 ______ *(b) 280* ______ *(c) 70* ______ *(d) 14* ______

4. Tiles are 7 cm long. How many can you fit into a space:

(a) 42 cm long? ______ *tiles* *(b) 35 cm long?* ______ *tiles* *(c) 63 cm long?* ______ *tiles*

5. Each bus can carry 70 passengers. How many buses would be needed to carry:

(a) 560 people? ______ *buses* *(b) 420 people?* ______ *buses* *(c) 210 people?* ______ *buses*

6. Complete these:

(a) 37 ÷ 7 = ______ *r* ______ *(b) 64 ÷ 7 =* ______ *r* ______ *(c) 25 ÷ 7 =* ______ *r* ______

7. If you cut these pizzas into sevenths, how many slices would you have?

(a) 4 pizzas ______ *slices* *(c) 7 pizzas* ______ *slices* *(e) 3 pizzas* ______ *slices*

(b) 5 pizzas ______ *slices* *(d) 9 pizzas* ______ *slices* *(f) 6 pizzas* ______ *slices*

8. A crate can hold 70 pears. How many pears would fit in:

(a) 3 crates? ______ *pears* *(b) 6 crates?* ______ *pears* *(c) 10 crates?* ______ *pears*

9. Stamps cost 70¢ each. Karl buys 5 and his friend Ahmed buys 8 stamps.

(a) How much money does Karl spend? ______________

(b) How much money does Ahmed spend? ______________

(c) How much more does Ahmed spend than Karl? ______________

(d) How much change would Karl receive from a $10 bill? ______________

(e) How much change would Ahmed receive from a $10 bill? ______________

10. Chutes 'n' Ladders

																				Finish
	81	82	83	84	85	86	87	88	89	90	91	92	93	94	95	96	97	98	99	100
	80	79	78	77	76	75	74	73	72	71	70	69	68	67	66	65	64	63	62	61
	41	42	43	44	45	46	47	48	49	50	51	52	53	54	55	56	57	58	59	60
	40	39	38	37	36	35	34	33	32	31	30	29	28	27	26	25	24	23	22	21
Start	1	2	3	4	5	6	7	8	9	10	11	12	13	14	15	16	17	18	19	20

www.worldteacherspress.com ©World Teachers Press®

Set A

1. 7 + 14 = ____	21. 7 + 21 = ____
2. 7 x 6 = ____	22. 7 x 2 = ____
3. 21 ÷ 7 = ____	23. 28 ÷ 7 = ____
4. 14 – 7 = ____	24. 49 – 7 = ____
5. 7 + 28 = ____	25. 7 + 42 = ____
6. 7 x 3 = ____	26. 7 x 5 = ____
7. 49 ÷ 7 = ____	27. 42 ÷ 7 = ____
8. 70 – 7 = ____	28. 77 – 7 = ____
9. 35 + 7 = ____	29. 56 + 7 = ____
10. 7 x 7 = ____	30. 7 x 9 = ____
11. 14 ÷ 7 = ____	31. 70 ÷ 7 = ____
12. 21 – 7 = ____	32. 42 – 7 = ____
13. 49 + 7 = ____	33. 7 x 70 = ____
14. 7 x 0 = ____	34. 7 x 10 = ____
15. 35 ÷ 7 = ____	35. 7 ÷ 7 = ____
16. 35 – 7 = ____	36. 63 – 7 = ____
17. 7 + 70 = ____	37. 7 + 63 = ____
18. 7 x 70 = ____	38. 7 x 4 = ____
19. 63 ÷ 7 = ____	39. 56 ÷ 7 = ____
20. 28 – 7 = ____	40. 70 x 0 = ____

My score:

My time:

____ mins ____ secs

I'm happy · I'm not happy · OOPS! · I didn't understand

The main area for me to work on is:

____

____

____

____

____

____

____

Set B

1. 7 + 7 = ____	21. 7 + 56 = ____
2. 7 x 8 = ____	22. 3 x 70 = ____
3. 42 ÷ 7 = ____	23. 28 ÷ 7 = ____
4. 70 – 7 = ____	24. 35 – 7 = ____
5. 21 + 7 = ____	25. 7 + 49 = ____
6. 7 x 5 = ____	26. 7 x 4 = ____
7. 7 ÷ 7 = ____	27. 49 ÷ 7 = ____
8. 28 – 7 = ____	28. 49 – 7 = ____
9. 7 + 63 = ____	29. 7 + 35 = ____
10. 7 x 3 = ____	30. 7 x 6 = ____
11. 35 ÷ 7 = ____	31. 56 ÷ 7 = ____
12. 14 – 7 = ____	32. 63 – 7 = ____
13. 7 – 7 = ____	33. 28 + 7 = ____
14. 7 x 9 = ____	34. 4 x 70 = ____
15. 14 ÷ 7 = ____	35. 21 ÷ 7 = ____
16. 2 x 70 = ____	36. 56 – 7 = ____
17. 42 + 7 = ____	37. 14 + 7 = ____
18. 7 x 7 = ____	38. 7 x 10 = ____
19. 7 x 0 = ____	39. 56 ÷ 7 = ____
20. 21 – 7 = ____	40. 42 – 7 = ____

My score:

My time:

____ mins ____ secs

I'm happy · I'm not happy · OOPS! · I didn't understand

The main area for me to work on is:

____

____

____

____

____

____

____

©World Teachers Press® www.worldteacherspress.com

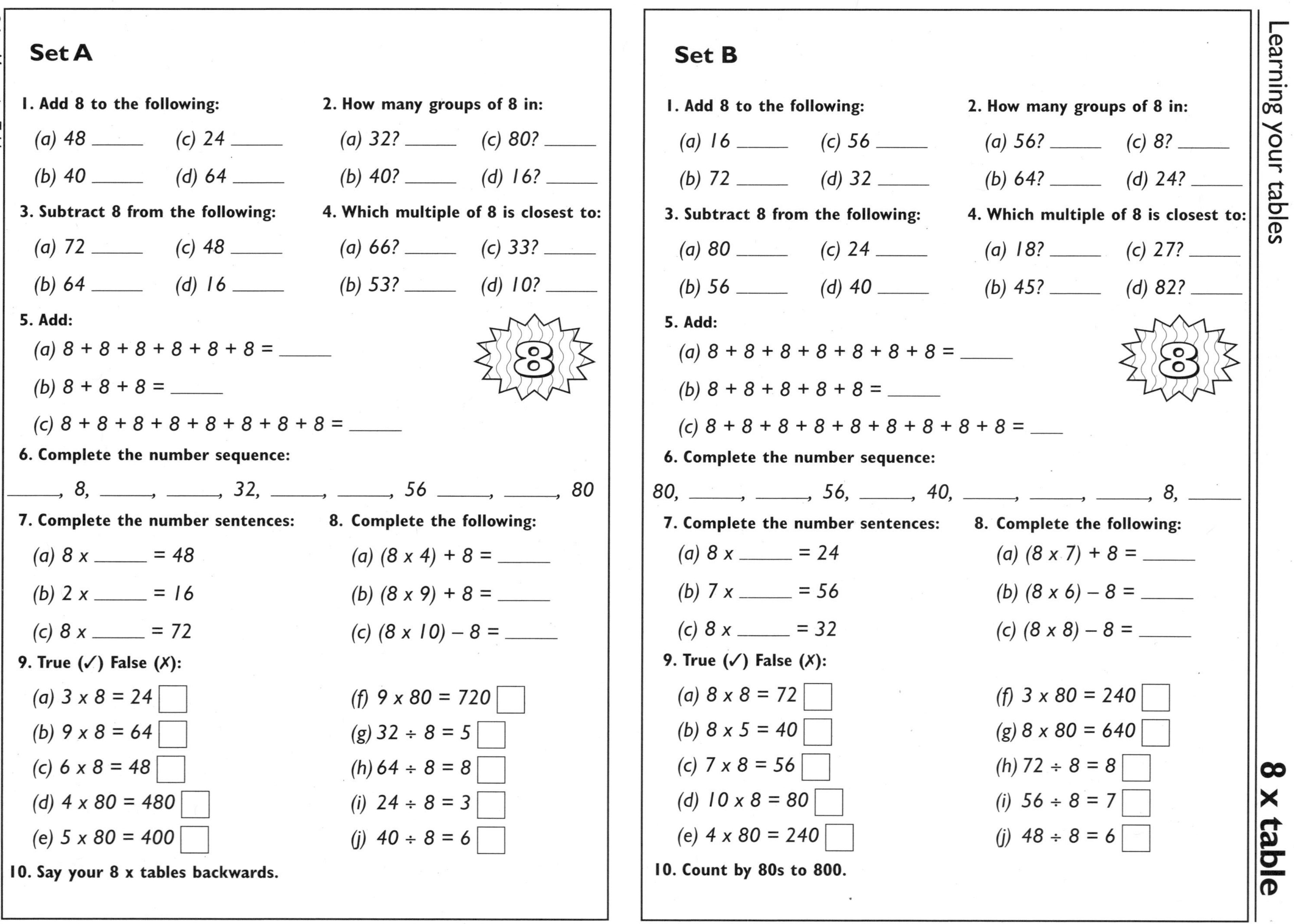

Set A

1. Add 8 to the following:

(a) 48 ______ (c) 24 ______

(b) 40 ______ (d) 64 ______

2. How many groups of 8 in:

(a) 32? ______ (c) 80? ______

(b) 40? ______ (d) 16? ______

3. Subtract 8 from the following:

(a) 72 ______ (c) 48 ______

(b) 64 ______ (d) 16 ______

4. Which multiple of 8 is closest to:

(a) 66? ______ (c) 33? ______

(b) 53? ______ (d) 10? ______

5. Add:

(a) 8 + 8 + 8 + 8 + 8 + 8 = ______

(b) 8 + 8 + 8 = ______

(c) 8 + 8 + 8 + 8 + 8 + 8 + 8 + 8 = ______

6. Complete the number sequence:

______, 8, ______, ______, 32, ______, ______, 56 ______, ______, 80

7. Complete the number sentences:

(a) 8 x ______ = 48

(b) 2 x ______ = 16

(c) 8 x ______ = 72

8. Complete the following:

(a) (8 x 4) + 8 = ______

(b) (8 x 9) + 8 = ______

(c) (8 x 10) – 8 = ______

9. True (✓) False (✗):

(a) 3 x 8 = 24 ☐

(b) 9 x 8 = 64 ☐

(c) 6 x 8 = 48 ☐

(d) 4 x 80 = 480 ☐

(e) 5 x 80 = 400 ☐

(f) 9 x 80 = 720 ☐

(g) 32 ÷ 8 = 5 ☐

(h) 64 ÷ 8 = 8 ☐

(i) 24 ÷ 8 = 3 ☐

(j) 40 ÷ 8 = 6 ☐

10. Say your 8 x tables backwards.

Set B

1. Add 8 to the following:

(a) 16 ______ (c) 56 ______

(b) 72 ______ (d) 32 ______

2. How many groups of 8 in:

(a) 56? ______ (c) 8? ______

(b) 64? ______ (d) 24? ______

3. Subtract 8 from the following:

(a) 80 ______ (c) 24 ______

(b) 56 ______ (d) 40 ______

4. Which multiple of 8 is closest to:

(a) 18? ______ (c) 27? ______

(b) 45? ______ (d) 82? ______

5. Add:

(a) 8 + 8 + 8 + 8 + 8 + 8 + 8 = ______

(b) 8 + 8 + 8 + 8 + 8 = ______

(c) 8 + 8 + 8 + 8 + 8 + 8 + 8 + 8 + 8 = ___

6. Complete the number sequence:

80, ______, ______, 56, ______, 40, ______, ______, ______, 8, ______

7. Complete the number sentences:

(a) 8 x ______ = 24

(b) 7 x ______ = 56

(c) 8 x ______ = 32

8. Complete the following:

(a) (8 x 7) + 8 = ______

(b) (8 x 6) – 8 = ______

(c) (8 x 8) – 8 = ______

9. True (✓) False (✗):

(a) 8 x 8 = 72 ☐

(b) 8 x 5 = 40 ☐

(c) 7 x 8 = 56 ☐

(d) 10 x 8 = 80 ☐

(e) 4 x 80 = 240 ☐

(f) 3 x 80 = 240 ☐

(g) 8 x 80 = 640 ☐

(h) 72 ÷ 8 = 8 ☐

(i) 56 ÷ 8 = 7 ☐

(j) 48 ÷ 8 = 6 ☐

10. Count by 80s to 800.

www.worldteacherspress.com ©World Teachers Press®

0 x 8 = ____	0 x 80 = ____		
1 x 8 = ____	1 x 80 = ____	6 x 8 = ____	6 x 80 = ____
2 x 8 = ____	2 x 80 = ____	7 x 8 = ____	7 x 80 = ____
3 x 8 = ____	3 x 80 = ____	8 x 8 = ____	8 x 80 = ____
4 x 8 = ____	4 x 80 = ____	9 x 8 = ____	9 x 80 = ____
5 x 8 = ____	5 x 80 = ____	10 x 8 = ____	10 x 80 = ____

1. How many days in:

(a) 8 weeks 4 days? ____________ (b) 8 weeks 2 days? ____________

2. Double:

(a) 160 ____ (b) 240 ____ (c) 80 ____ (d) 8 ____

3. Halve:

(a) 64 ____ (b) 48 ____ (c) 16 ____ (d) 80 ____

4. Tiles are 8 cm long. How many can you fit into a space:

(a) 40 cm long? ____ tiles (b) 80 cm long? ____ tiles (c) 56 cm long? ____ tiles

5. Each train car can carry 80 passengers. How many cars would be needed to carry:

(a) 320 people? ____ cars (b) 640 people? ____ cars (c) 160 people? ____ cars

6. Complete these:

(a) 74 ÷ 8 = ____ r ____ (b) 35 ÷ 8 = ____ r ____ (c) 21 ÷ 8 = ____ r ____

7. If you cut these apples pies into eighths, how many slices would you have?

(a) 3 pies ____ slices (c) 9 pies ____ slices (e) 5 pies ____ slices

(b) 6 pies ____ slices (d) 4 pies ____ slices (f) 10 pies ____ slices

8. A crate can hold 80 bottles. How many bottles would fit in:

(a) 5 crates? ____ bottles (b) 9 crates? ____ bottles (c) 7 crates? ____ bottles

9. Tennis balls cost 80¢ each. Sarah buys 4 and her sister Leah buys 6 tennis balls.

(a) How much money does Sarah spend? ____________

(b) How much money does Leah spend? ____________

(c) How much do they spend altogether? ____________

(d) How much change would Sarah receive from a $10 bill? ____________

(e) How much change would Leah receive from a $10 bill? ____________

10. Chutes 'n' Ladders

Finish

81	82	83	84	85	86	87	88	89	90	91	92	93	94	95	96	97	98	99	100
80	79	78	77	76	75	74	73	72	71	70	69	68	67	66	65	64	63	62	61
41	42	43	44	45	46	47	48	49	50	51	52	53	54	55	56	57	58	59	60
40	39	38	37	36	35	34	33	32	31	30	29	28	27	26	25	24	23	22	21
1	2	3	4	5	6	7	8	9	10	11	12	13	14	15	16	17	18	19	20

Start

©World Teachers Press® www.worldteacherspress.com

Set A

1. 16 + 8 = ____
2. 6 x 8 = ____
3. 24 ÷ 8 = ____
4. 16 – 8 = ____
5. 8 + 32 = ____
6. 8 x 3 = ____
7. 56 ÷ 8 = ____
8. 80 – 8 = ____
9. 8 + 40 = ____
10. 8 x 8 = ____
11. 16 ÷ 8 = ____
12. 24 – 8 = ____
13. 56 + 8 = ____
14. 8 x 0 = ____
15. 40 ÷ 8 = ____
16. 40 – 8 = ____
17. 8 + 80 = ____
18. 8 x 80 = ____
19. 64 ÷ 8 = ____
20. 32 – 8 = ____
21. 24 + 8 = ____
22. 8 x 2 = ____
23. 32 ÷ 8 = ____
24. 56 – 8 = ____
25. 8 + 48 = ____
26. 8 x 5 = ____
27. 48 ÷ 8 = ____
28. 88 – 8 = ____
29. 64 + 8 = ____
30. 8 x 9 = ____
31. 80 ÷ 8 = ____
32. 48 – 8 = ____
33. 3 x 80 = ____
34. 8 x 10 = ____
35. 8 ÷ 8 = ____
36. 72 – 8 = ____
37. 8 + 72 = ____
38. 8 x 4 = ____
39. 72 ÷ 8 = ____
40. 2 x 80 = ____

My score: ☐

My time:

____ *mins* ____ *secs*

I'm happy I'm not happy OOPS! I didn't understand

The main area for me to work on is:

____
____
____
____
____
____
____

Set B

1. 80 + 8 = ____
2. 8 x 8 = ____
3. 48 ÷ 8 = ____
4. 80 – 8 = ____
5. 24 + 8 = ____
6. 8 x 5 = ____
7. 8 ÷ 8 = ____
8. 32 – 8 = ____
9. 8 + 72 = ____
10. 8 x 3 = ____
11. 40 ÷ 8 = ____
12. 16 – 8 = ____
13. 8 + 8 = ____
14. 9 x 8 = ____
15. 16 ÷ 8 = ____
16. 88 – 8 = ____
17. 48 + 8 = ____
18. 8 x 7 = ____
19. 2 x 80 = ____
20. 24 – 8 = ____
21. 72 + 8 = ____
22. 80 x 0 = ____
23. 32 ÷ 8 = ____
24. 40 – 8 = ____
25. 8 + 56 = ____
26. 8 x 4 = ____
27. 56 ÷ 8 = ____
28. 56 – 8 = ____
29. 8 + 40 = ____
30. 6 x 8 = ____
31. 64 ÷ 8 = ____
32. 72 – 8 = ____
33. 8 + 32 = ____
34. 5 x 80 = ____
35. 24 ÷ 8 = ____
36. 64 – 8 = ____
37. 16 + 8 = ____
38. 10 x 8 = ____
39. 72 ÷ 8 = ____
40. 48 – 8 = ____

My score: ☐

My time:

____ *mins* ____ *secs*

I'm happy I'm not happy OOPS! I didn't understand

The main area for me to work on is:

____
____
____
____
____
____
____

www.worldteacherspress.com ©World Teachers Press®

Set A

1. Add 9 to the following:

(a) 27 ______ (c) 72 ______

(b) 45 ______ (d) 18 ______

2. How many groups of 9 in:

(a) 90? ______ (c) 36? ______

(b) 63? ______ (d) 81? ______

3. Subtract 9 from the following:

(a) 18 ______ (c) 36 ______

(b) 81 ______ (d) 54 ______

4. Which multiple of 9 is closest to:

(a) 83? ______ (c) 50? ______

(b) 21? ______ (d) 30? ______

5. Add:

(a) 9 + 9 + 9 + 9 + 9 = ______

(b) 9 + 9 + 9 = ______

(c) 9 + 9 + 9 + 9 + 9 + 9 + 9 + 9 = ______

6. Complete the number sequence:

0, 9, ______, ______, 36, ______, ______, 63, ______, ______, 90

7. Complete the number sentences:

(a) 9 x ______ = 45

(b) 3 x ______ = 27

(c) 9 x ______ = 72

8. Complete the following:

(a) (9 x 3) + 9 = ______

(b) (9 x 5) + 9 = ______

(c) (9 x 6) – 9 = ______

9. True (✓) False (✗):

(a) 3 x 9 = 27 ☐

(b) 9 x 9 = 81 ☐

(c) 6 x 9 = 54 ☐

(d) 8 x 9 = 63 ☐

(e) 3 x 90 = 270 ☐

(f) 6 x 90 = 630 ☐

(g) 8 x 90 = 720 ☐

(h) 54 ÷ 9 = 6 ☐

(i) 63 ÷ 9 = 7 ☐

(j) 81 ÷ 9 = 8 ☐

10. Say your 9 x tables backwards.

Set B

1. Add 9 to the following:

(a) 54 ______ (c) 81 ______

(b) 9 ______ (d) 63 ______

2. How many groups of 9 in:

(a) 18? ______ (c) 54? ______

(b) 72? ______ (d) 27? ______

3. Subtract 9 from the following:

(a) 63 ______ (c) 90 ______

(b) 27 ______ (d) 45 ______

4. Which multiple of 9 is closest to:

(a) 35? ______ (c) 65? ______

(b) 70? ______ (d) 89? ______

5. Add:

(a) 9 + 9 + 9 + 9 + 9 + 9 + 9 = ______

(b) 9 + 9 + 9 + 9 = ______

(c) 9 + 9 + 9 + 9 + 9 + 9 + 9 + 9 + 9 = ______

6. Complete the number sequence:

______, ______, 72, ______, ______, 45, ______, ______, 18, ______, ______

7. Complete the number sentences:

(a) 9 x ______ = 36

(b) 9 x ______ = 0

(c) 9 x ______ = 81

8. Complete the following:

(a) (9 x 9) + 9 = ______

(b) (9 x 10) – 9 = ______

(c) (9 x 2) – 9 = ______

9. True (✓) False (✗):

(a) 5 x 9 = 36 ☐

(b) 7 x 9 = 72 ☐

(c) 10 x 9 = 90 ☐

(d) 7 x 90 = 630 ☐

(e) 5 x 90 = 450 ☐

(f) 9 x 90 = 810 ☐

(g) 4 x 90 = 270 ☐

(h) 45 ÷ 9 = 5 ☐

(i) 27 ÷ 9 = 4 ☐

(j) 72 ÷ 9 = 8 ☐

10. Count by 90s to 900.

©World Teachers Press® www.worldteacherspress.com

0 x 9 = ____	0 x 90 = ____		
1 x 9 = ____	1 x 90 = ____	6 x 9 = ____	6 x 90 = ____
2 x 9 = ____	2 x 90 = ____	7 x 9 = ____	7 x 90 = ____
3 x 9 = ____	3 x 90 = ____	8 x 9 = ____	8 x 90 = ____
4 x 9 = ____	4 x 90 = ____	9 x 9 = ____	9 x 90 = ____
5 x 9 = ____	5 x 90 = ____	10 x 9 = ____	10 x 90 = ____

1. How many days in:

(a) 9 weeks 3 days? ____________ *(b) 9 weeks 5 days? ____________*

2. Double:

(a) 180 ____ *(b) 450 ____* *(c) 90 ____* *(d) 18 ____*

3. Halve:

(a) 180 ____ *(b) 54 ____* *(c) 18 ____* *(d) 90 ____*

4. Tiles are 9 cm long. How many can you fit into a space:

(a) 45 cm long? ____ tiles *(b) 90 cm long? ____ tiles* *(c) 72 cm long? ____ tiles*

5. Each plane can carry 90 passengers. How many planes would be needed to carry:

(a) 270 people? ____ planes *(b) 540 people? ____ planes* *(c) 810 people? ____ planes*

6. Complete these:

(a) 38 ÷ 9 = ____ r ____ *(b) 59 ÷ 9 = ____ r ____* *(c) 79 ÷ 9 = ____ r ____*

7. If you cut these cakes into ninths, how many pieces would you have?

(a) 4 cakes ____ slices *(c) 6 cakes ____ slices* *(e) 2 cakes ____ slices*

(b) 9 cakes ____ slices *(d) 5 cakes ____ slices* *(f) 7 cakes ____ slices*

8. A pouch can hold 90 marbles. How many marbles would fit in:

(a) 7 pouches? ____ marbles *(b) 9 pouches? ____ marbles* *(c) 3 pouches? ____ marbles*

9. Comic books cost 90¢ each. Hanna buys 8 and her friend Toshia buys 4 comic books.

(a) How much money does Hanna spend? ____________

(b) How much money does Toshia spend? ____________

(c) How much more does Hanna spend than Toshia? ____________

(d) How much change would Hanna receive from a $10 bill? ____________

(e) How much change would Toshia receive from a $10 bill? ____________

10. **Chutes 'n' Ladders**

Finish

81	82	83	84	85	86	87	88	89	90	91	92	93	94	95	96	97	98	99	100
80	79	78	77	76	75	74	73	72	71	70	69	68	67	66	65	64	63	62	61
41	42	43	44	45	46	47	48	49	50	51	52	53	54	55	56	57	58	59	60
40	39	38	37	36	35	34	33	32	31	30	29	28	27	26	25	24	23	22	21
1	2	3	4	5	6	7	8	9	10	11	12	13	14	15	16	17	18	19	20

Start

www.worldteacherspress.com ©World Teachers Press®

Set A

1. 9 + 18 = ____	21. 27 + 9 = ____
2. 6 x 9 = ____	22. 9 x 2 = ____
3. 27 ÷ 9 = ____	23. 36 ÷ 9 = ____
4. 18 – 9 = ____	24. 63 – 9 = ____
5. 36 + 9 = ____	25. 54 + 9 = ____
6. 9 x 3 = ____	26. 9 x 5 = ____
7. 63 ÷ 9 = ____	27. 54 ÷ 9 = ____
8. 45 – 9 = ____	28. 99 – 9 = ____
9. 45 + 9 = ____	29. 72 + 9 = ____
10. 9 x 7 = ____	30. 9 x 9 = ____
11. 18 ÷ 9 = ____	31. 90 ÷ 9 = ____
12. 27 – 9 = ____	32. 81 – 9 = ____
13. 63 + 9 = ____	33. 9 x 90 = ____
14. 9 x 0 = ____	34. 9 x 10 = ____
15. 45 ÷ 9 = ____	35. 9 ÷ 9 = ____
16. 90 – 9 = ____	36. 54 – 9 = ____
17. 9 + 90 = ____	37. 9 + 81 = ____
18. 9 x 1 = ____	38. 9 x 4 = ____
19. 81 ÷ 9 = ____	39. 72 ÷ 9 = ____
20. 36 – 9 = ____	40. 4 x 90 = ____

My score: ☐

My time:

____ mins ____ secs

I'm happy

I'm not happy

OOPS!

I didn't understand

The main area for me to work on is:

____

____

____

____

____

____

____

Set B

1. 27 + 9 = ____	21. 9 + 45 = ____
2. 9 x 10 = ____	22. 9 x 0 = ____
3. 18 ÷ 2 = ____	23. 27 ÷ 9 = ____
4. 81 – 9 = ____	24. 36 – 9 = ____
5. 9 + 54 = ____	25. 2 x 90 = ____
6. 9 x 4 = ____	26. 3 x 90 = ____
7. 45 ÷ 9 = ____	27. 63 ÷ 9 = ____
8. 54 – 9 = ____	28. 45 – 9 = ____
9. 72 + 9 = ____	29. 9 + 81 = ____
10. 9 x 9 = ____	30. 9 x 7 = ____
11. 81 ÷ 9 = ____	31. 90 ÷ 9 = ____
12. 90 x 0 = ____	32. 18 – 9 = ____
13. 9 + 18 = ____	33. 63 + 9 = ____
14. 9 x 5 = ____	34. 9 x 3 = ____
15. 36 ÷ 9 = ____	35. 72 ÷ 9 = ____
16. 27 – 9 = ____	36. 63 – 9 = ____
17. 36 + 9 = ____	37. 9 + 90 = ____
18. 9 x 2 = ____	38. 6 x 9 = ____
19. 54 ÷ 9 = ____	39. 9 ÷ 9 = ____
20. 90 – 9 = ____	40. 99 – 9 = ____

My score: ☐

My time:

____ mins ____ secs

I'm happy

I'm not happy

OOPS!

I didn't understand

The main area for me to work on is:

____

____

____

____

____

____

____

©World Teachers Press® www.worldteacherspress.com

Set A

1. Add 10 to the following:

(a) 48 ______ (c) 20 ______

(b) 50 ______ (d) 90 ______

2. How many groups of 10 in:

(a) 30? ______ (c) 80? ______

(b) 40? ______ (d) 110? ______

3. Subtract 10 from the following:

(a) 50 ______ (c) 90 ______

(b) 20 ______ (d) 40 ______

4. Which multiple of 10 is closest to:

(a) 57? ______ (c) 48? ______

(b) 33? ______ (d) 87? ______

5. Add:

(a) 10 + 10 + 10 + 10 = ______

(b) 10 + 10 + 10 + 10 + 10 + 10 = ______

(c) 10 + 10 + 10 + 10 + 10 + 10 + 10 + 10 = ______

6. Complete the number sequence:

0, ______, ______, ______, **40,** ______, ______, **70** ______, ______, **100**

7. Complete the number sentences:

(a) 10 x ______ = 70

(b) 10 x ______ = 100

(c) 10 x ______ = 50

8. Complete the following:

(a) (10 x 10) + 10 = ______

(b) (10 x 6) + 10 = ______

(c) (10 x 10) – 10 = ______

9. True (✓) False (✗):

(a) 3 x 10 = 30 ☐

(b) 9 x 10 = 90 ☐

(c) 6 x 10 = 70 ☐

(d) 7 x 100 = 700 ☐

(e) 5 x 100 = 500 ☐

(f) 9 x 100 = 800 ☐

(g) 90 ÷ 10 = 8 ☐

(h) 70 ÷ 10 = 7 ☐

(i) 60 ÷ 10 = 6 ☐

(j) 30 ÷ 10 = 2 ☐

10. Say your 10 x tables backwards.

Set B

1. Add 10 to the following:

(a) 10 ______ (c) 80 ______

(b) 70 ______ (d) 110 ______

2. How many groups of 10 in:

(a) 50? ______ (c) 90? ______

(b) 60? ______ (d) 200? ______

3. Subtract 10 from the following:

(a) 80 ______ (c) 60 ______

(b) 110 ______ (d) 100 ______

4. Which multiple of 10 is closest to:

(a) 18? ______ (c) 27? ______

(b) 74? ______ (d) 98? ______

5. Add:

(a) 10 + 10 + 10 = ______

(b) 10 + 10 + 10 + 10 + 10 + 10 + 10 = ______

(c) 10 + 10 + 10 + 10 + 10 + 10 + 10 + 10 + 10 = ______

6. Complete the number sequence:

100, ______, ______, 70, ______, 50, ______, ______, ______, 10, ______

7. Complete the number sentences:

(a) 10 x ______ = 20

(b) 8 x ______ = 80

(c) 10 x ______ = 60

8. Complete the following:

(a) (10 x 4) + 10 = ______

(b) (10 x 7) – 10 = ______

(c) (10 x 5) – 10 = ______

9. True (✓) False (✗):

(a) 8 x 10 = 80 ☐

(b) 5 x 10 = 40 ☐

(c) 7 x 10 = 60 ☐

(d) 10 x 10 = 100 ☐

(e) 4 x 100 = 400 ☐

(f) 3 x 100 = 500 ☐

(g) 6 x 100 = 600 ☐

(h) 40 ÷ 10 = 5 ☐

(i) 80 ÷ 10 = 8 ☐

(j) 50 ÷ 10 = 6 ☐

10. Count by 100s to 1000.

www.worldteacherspress.com ©World Teachers Press®

0 x 10 = ____	*0 x 100 = ____*		
1 x 10 = ____	*1 x 100 = ____*	*6 x 10 = ____*	*6 x 100 = ____*
2 x 10 = ____	*2 x 100 = ____*	*7 x 10 = ____*	*7 x 100 = ____*
3 x 10 = ____	*3 x 100 = ____*	*8 x 10 = ____*	*8 x 100 = ____*
4 x 10 = ____	*4 x 100 = ____*	*9 x 10 = ____*	*9 x 100 = ____*
5 x 10 = ____	*5 x 100 = ____*	*10 x 10 = ____*	*10 x 100 = ____*

1. How many days in:

(a) 10 weeks 4 days? ____________ *(b) 10 weeks 1 day? ____________*

2. Double:

(a) 500 ____ *(b) 300 ____* *(c) 40 ____* *(d) 20 ____*

3. Halve:

(a) 20 ____ *(b) 60 ____* *(c) 800 ____* *(d) 100 ____*

4. Candies cost 10¢ each. How many can you buy with:

(a) $1.50? ____ *(b) $2.70? ____* *(c) $1.20? ____* *(d) $6.00? ____*

5. Each bus can carry 100 passengers. How many buses would be needed to carry:

(a) 800 people? ____ buses *(b) 1000 people? ____ buses* *(c) 200 people? ____ buses*

6. Complete these:

(a) 72 ÷ 10 = ____ r ____ *(b) 67 ÷ 10 = ____ r ____* *(c) 103 ÷ 10 = ____ r ____*

7. If you cut these pizzas into tenths, how many slices would you have?

(a) 10 pizzas ____ slices *(b) 7 pizzas ____ slices* *(c) 2 pizzas ____ slices*

(d) 4 pizzas ____ slices *(e) 15 pizzas ____ slices* *(f) 20 pizzas ____ slices*

8. A crate can hold 100 bottles. How many bottles would fit in:

(a) 6 crates? ____ bottles *(b) 10 crates? ____ bottles* *(c) 20 crates? ____ bottles*

9. Notepads cost $1.00 each. Jamie buys 8 and Nikki buys 4 pads.

(a) How much money does Jamie spend? ____________

(b) How much money does Nikki spend? ____________

(c) How much do Jamie and Nikki spend altogether? ____________

(d) How much change would Jamie receive from a $20 bill? ____________

(e) How much change would Nikki receive from a $20 bill? ____________

10. Chutes 'n' Ladders

																			Finish
81	82	83	84	85	86	87	88	89	90	91	92	93	94	95	96	97	98	99	100
80	79	78	77	76	75	74	73	72	71	70	69	68	67	66	65	64	63	62	61
41	42	43	44	45	46	47	48	49	50	51	52	53	54	55	56	57	58	59	60
40	39	38	37	36	35	34	33	32	31	30	29	28	27	26	25	24	23	22	21
Start 1	2	3	4	5	6	7	8	9	10	11	12	13	14	15	16	17	18	19	20

©World Teachers Press® www.worldteacherspress.com

Set A

1. 10 + 30 = ____
2. 10 x 8 = ____
3. 70 ÷ 10 = ____
4. 500 – 10 = ____
5. 10 + 90 = ____
6. 10 x 10 = ____
7. 40 ÷ 10 = ____
8. 800 – 10 = ____
9. 10 + 50 = ____
10. 10 x 6 = ____
11. 100 ÷ 10 = ____
12. 100 – 10 = ____
13. 120 + 10 = ____
14. 10 x 0 = ____
15. 50 ÷ 10 = ____
16. 300 – 10 = ____
17. 10 + 40 = ____
18. 20 x 10 = ____
19. 300 ÷ 10 = ____
20. 900 – 10 = ____
21. 100 + 70 = ____
22. 8 x 10 = ____
23. 150 ÷ 10 = ____
24. 110 – 10 = ____
25. 10 + 20 = ____
26. 2 x 10 = ____
27. 200 ÷ 10 = ____
28. 120 – 10 = ____
29. 100 + 10 = ____
30. 6 x 10 = ____
31. 40 ÷ 10 = ____
32. 20 – 10 = ____
33. 10 + 200 = ____
34. 10 x 100 = ____
35. 70 ÷ 10 = ____
36. 40 – 10 = ____
37. 10 + 60 = ____
38. 10 x 9 = ____
39. 60 ÷ 10 = ____
40. 150 – 10 = ____

My score: ☐

My time:

____ *mins* ____ *secs*

I'm happy · I'm not happy · OOPS! · I didn't understand

The main area for me to work on is:

____

Set B

1. 10 + 80 = ____
2. 4 x 10 = ____
3. 100 ÷ 10 = ____
4. 50 – 10 = ____
5. 40 + 10 = ____
6. 9 x 10 = ____
7. 150 ÷ 10 = ____
8. 90 – 10 = ____
9. 10 + 70 = ____
10. 10 x 10 = ____
11. 700 ÷ 10 = ____
12. 10 – 10 = ____
13. 10 + 90 = ____
14. 70 x 10 = ____
15. 400 ÷ 10 = ____
16. 100 – 10 = ____
17. 10 + 100 = ____
18. 20 x 10 = ____
19. 60 x 10 = ____
20. 120 – 10 = ____
21. 100 + 60 = ____
22. 50 x 10 = ____
23. 200 ÷ 10 = ____
24. 150 – 10 = ____
25. 10 + 20 = ____
26. 30 x 10 = ____
27. 600 ÷ 10 = ____
28. 200 – 10 = ____
29. 10 + 140 = ____
30. 10 x 0 = ____
31. 800 ÷ 10 = ____
32. 170 – 10 = ____
33. 200 + 10 = ____
34. 4 x 100 = ____
35. 500 ÷ 10 = ____
36. 300 – 10 = ____
37. 190 + 10 = ____
38. 9 x 10 = ____
39. 300 ÷ 10 = ____
40. 250 – 10 = ____

My score: ☐

My time:

____ *mins* ____ *secs*

I'm happy · I'm not happy · OOPS! · I didn't understand

The main area for me to work on is:

____

www.worldteacherspress.com ©World Teachers Press®

Set A

1. Add 12 to the following:

(a) 36 ______ (c) 60 ______

(b) 96 ______ (d) 24 ______

2. How many groups of 12 in:

(a) 24? ______ (c) 72? ______

(b) 108? ______ (d) 84? ______

3. Subtract 12 from the following:

(a) 48 ______ (c) 108 ______

(b) 24 ______ (d) 732 ______

4. Which multiple of 12 is closest to:

(a) 38? ______ (c) 80? ______

(b) 57? ______ (d) 94? ______

5. Add:

(a) 12 + 12 + 12 = ______

(b) 12 + 12 + 12 + 12 + 12 + 12 + 12 = ______

(c) 12 + 12 + 12 + 12 + 12 = ______

6. Complete the number sequence:

0, ______, 24, ______, 48, ______, ______, 84, ______, ______, 120

7. Complete the number sentences:

(a) 12 x ______ = 24

(b) 12 x ______ = 120

(c) 12 x ______ = 72

8. Complete the following:

(a) (12 x 6) + 12 = ______

(b) (12 x 10) – 12 = ______

(c) (12 x 4) + 12 = ______

9. True (✓) False (✗):

(a) 3 x 12 = 36 ☐

(b) 9 x 12 = 96 ☐

(c) 6 x 12 = 72 ☐

(d) 8 x 12 = 84 ☐

(e) 7 x 12 = 840 ☐

(f) 5 x 120 = 600 ☐

(g) 2 x 120 = 240 ☐

(h) 108 ÷ 12 = 8 ☐

(i) 84 ÷ 12 = 7 ☐

(j) 72 ÷ 12 = 6 ☐

10. Say your 12 x tables.

Set B

1. Add 12 to the following:

(a) 12 ______ (c) 48 ______

(b) 84 ______ (d) 108 ______

2. How many groups of 12 in:

(a) 60? ______ (c) 36? ______

(b) 120? ______ (d) 48? ______

3. Subtract 12 from the following:

(a) 60 ______ (c) 120 ______

(b) 84 ______ (d) 36 ______

4. Which multiple of 12 is closest to:

(a) 100? ______ (c) 25? ______

(b) 42? ______ (d) 63? ______

5. Add:

(a) 12 + 12 + 12 + 12 = ______

(b) 12 + 12 + 12 + 12 + 12 + 12 = ______

(c) 12 + 12 + 12 + 12 + 12 + 12 + 12 + 12 + 12 = ______

6. Complete the number sequence:

120, ______, ______, 84, ______, 60, ______, ______, ______, 12, ______

7. Complete the number sentences:

(a) 12 x ______ = 108

(b) 12 x ______ = 0

(c) 12 x ______ = 96

8. Complete the following:

(a) (12 x 7) + 12 = ______

(b) (12 x 5) – 12 = ______

(c) (12 x 8) – 12 = ______

9. True (✓) False (✗):

(a) 5 x 12 = 60 ☐

(b) 7 x 12 = 72 ☐

(c) 10 x 12 = 120 ☐

(d) 4 x 120 = 600 ☐

(e) 3 x 120 = 360 ☐

(f) 6 x 120 = 720 ☐

(g) 48 ÷ 12 = 5 ☐

(h) 96 ÷ 12 = 8 ☐

(i) 36 ÷ 12 = 3 ☐

(j) 60 ÷ 12 = 6 ☐

10. Count by 12s to 120.

©World Teachers Press® www.worldteacherspress.com

0 x 12 = _____	0 x 120 = _____		
1 x 12 = _____	1 x 120 = _____	6 x 12 = _____	6 x 120 = _____
2 x 12 = _____	2 x 120 = _____	7 x 12 = _____	7 x 120 = _____
3 x 12 = _____	3 x 120 = _____	8 x 12 = _____	8 x 120 = _____
4 x 12 = _____	4 x 120 = _____	9 x 12 = _____	9 x 120 = _____
5 x 12 = _____	5 x 120 = _____	10 x 12 = _____	10 x 120 = _____

1. How many days in:

(a) 12 weeks 3 days? _____________ *(b) 12 weeks 5 days?* _____________

2. Double:

(a) 60 _____ *(b) 48* _____ *(c) 40* _____ *(d) 20* _____

3. Halve:

(a) 24 _____ *(b) 120* _____ *(c) 20* _____ *(d) 60* _____

4. Tiles are 12 cm long. How many can you fit into a space:

(a) 60 cm long? _____ *tiles* *(b) 120 cm long?* _____ *tiles* *(c) 96 cm long?* _____ *tiles*

5. Each ferry can carry 120 passengers. How many ferries would be needed to carry:

(a) 360 people? _____ *ferries* *(b) 600 people?* _____ *ferries* *(c) 1200 people?* _____ *ferries*

6. Complete these:

(a) 50 ÷ 12 = _____ *r* _____ *(b) 41 ÷ 12 =* _____ *r* _____ *(c) 27 ÷ 12 =* _____ *r* _____

7. If you cut these oranges into twelfths, how many slices would you have?

(a) 2 oranges _____ *slices* *(b) 5 oranges* _____ *slices* *(c) 4 oranges* _____ *slices*

(d) 10 oranges _____ *slices* *(e) 7 oranges* _____ *slices* *(f) 8 oranges* _____ *slices*

8. A crate can hold 120 pineapples. How many pineapples would fit in:

(a) 3 crates? _____ *pineapples* *(b) 7 crates?* _____ *pineapples* *(c) 9 crates?* _____ *pineapples*

9. Gummy snakes cost 12¢ each. Emily buys 10 and her friend Anita buys 5 snakes.

(a) How much money does Emily spend? _____________

(b) How much money does Anita spend? _____________

(c) How much more does Emily spend than Anita? _____________

(d) How much change would Emily receive from a $5 bill? _____________

(e) How much change would Anita receive from a $5 bill? _____________

10. Tables Battleship.

	A	B	C	D
4				
3				
2				
1				

	A	B	C	D
4				
3				
2				
1				

	A	B	C	D
4				
3				
2				
1				

	A	B	C	D
4				
3				
2				
1				

www.worldteacherspress.com ©World Teachers Press®

Set A

1. 12 + 24 = ____
2. 6 x 12 = ____
3. 36 ÷ 12 = ____
4. 24 – 12 = ____
5. 48 + 12 = ____
6. 3 x 12 = ____
7. 84 ÷ 12 = ____
8. 60 – 12 = ____
9. 12 + 60 = ____
10. 12 x 0 = ____
11. 24 ÷ 12 = ____
12. 36 – 12 = ____
13. 84 + 12 = ____
14. 7 x 12 = ____
15. 60 ÷ 12 = ____
16. 120 – 12 = ____
17. 120 + 12 = ____
18. 12 x 1 = ____
19. 108 ÷ 12 = ____
20. 48 – 12 = ____
21. 12 + 36 = ____
22. 2 x 12 = ____
23. 48 ÷ 12 = ____
24. 84 – 12 = ____
25. 12 + 72 = ____
26. 5 x 12 = ____
27. 72 ÷ 12 = ____
28. 12 – 12 = ____
29. 96 + 12 = ____
30. 12 x 3 = ____
31. 120 ÷ 12 = ____
32. 108 – 12 = ____
33. 12 + 0 = ____
34. 10 x 12 = ____
35. 12 ÷ 12 = ____
36. 72 – 12 = ____
37. 108 + 12 = ____
38. 4 x 12 = ____
39. 96 ÷ 12 = ____
40. 120 x 0 = ____

My score: ☐

My time:

____ *mins* ____ *secs*

I'm happy
I'm not happy
OOPS!
I didn't understand

The main area for me to work on is:

____

____

____

____

____

____

____

Set B

1. 12 + 36 = ____
2. 10 x 12 = ____
3. 96 ÷ 12 = ____
4. 108 – 12 = ____
5. 72 + 12 = ____
6. 4 x 12 = ____
7. 72 ÷ 12 = ____
8. 72 – 12 = ____
9. 96 + 12 = ____
10. 12 x 0 = ____
11. 24 ÷ 12 = ____
12. 12 – 12 = ____
13. 12 + 12 = ____
14. 5 x 12 = ____
15. 48 ÷ 12 = ____
16. 84 – 12 = ____
17. 108 + 12 = ____
18. 2 x 12 = ____
19. 108 ÷ 12 = ____
20. 48 – 12 = ____
21. 12 + 24 = ____
22. 12 x 10 = ____
23. 12 ÷ 12 = ____
24. 120 – 12 = ____
25. 48 + 12 = ____
26. 9 x 12 = ____
27. 84 ÷ 12 = ____
28. 36 – 12 = ____
29. 12 + 60 = ____
30. 7 x 12 = ____
31. 36 ÷ 12 = ____
32. 60 – 12 = ____
33. 84 + 12 = ____
34. 12 x 3 = ____
35. 60 ÷ 12 = ____
36. 24 – 12 = ____
37. 120 + 12 = ____
38. 6 x 12 = ____
39. 120 x 0 = ____
40. 96 – 12 = ____

My score: ☐

My time:

____ *mins* ____ *secs*

I'm happy
I'm not happy
OOPS!
I didn't understand

The main area for me to work on is:

____

____

____

____

____

____

____

©World Teachers Press® www.worldteacherspress.com

Set A

1. Add 15 to the following:

(a) 75 ______ (c) 105 ______

(b) 45 ______ (d) 120 ______

2. How many groups of 15 in:

(a) 120? ______ (c) 15? ______

(b) 75? ______ (d) 105? ______

3. Subtract 15 from the following:

(a) 150 ______ (c) 45 ______

(b) 120 ______ (d) 60 ______

4. Which multiple of 15 is closest to:

(a) 151? ______ (c) 92? ______

(b) 33? ______ (d) 17? ______

5. Add:

(a) 15 + 15 + 15 = ______

(b) 15 + 15 + 15 + 15 + 15 = ______

(c) 15 + 15 + 15 + 15 + 15 + 15 + 15 + 15 = ______

6. Complete the number sequence:

0, 15, ______, ______, 60, ______, ______, 105, ______, ______, 150

7. Complete the number sentences:

(a) 15 x ______ = 45

(b) 15 x ______ = 60

(c) 15 x ______ = 90

8. Complete the following:

(a) (15 x 3) + 15 = ______

(b) (15 x 7) + 15 = ______

(c) (15 x 6) – 15 = ______

9. True (✓) False (✗):

(a) 3 x 15 = 60 ☐

(b) 5 x 15 = 75 ☐

(c) 10 x 15 = 150 ☐

(d) 2 x 15 = 45 ☐

(e) 7 x 150 = 1050 ☐

(f) 2 x 150 = 300 ☐

(g) 4 x 150 = 600 ☐

(h) 75 ÷ 15 = 6 ☐

(i) 30 ÷ 15 = 2 ☐

(j) 150 ÷ 15 = 10 ☐

10. Say your 15 x tables.

Set B

1. Add 15 to the following:

(a) 135 ______ (c) 90 ______

(b) 60 ______ (d) 15 ______

2. How many groups of 15 in:

(a) 150? ______ (c) 30? ______

(b) 90? ______ (d) 45? ______

3. Subtract 15 from the following:

(a) 75 ______ (c) 90 ______

(b) 135 ______ (d) 30 ______

4. Which multiple of 15 is closest to:

(a) 100? ______ (c) 25? ______

(b) 42? ______ (d) 133? ______

5. Add:

(a) 15 + 15 + 15 + 15 = ______

(b) 15 + 15 + 15 + 15 + 15 + 15 = ______

(c) 15 + 15 + 15 + 15 + 15 + 15 + 15 + 15 + 15 = ______

6. Complete the number sequence:

150, ______, ______, 105, ______, 75, ______, ______, ______, 15, ______

7. Complete the number sentences:

(a) 15 x ______ = 150

(b) 15 x ______ = 75

(c) 15 x ______ = 30

8. Complete the following:

(a) (15 x 5) + 15 = ______

(b) (15 x 4) – 15 = ______

(c) (15 x 8) – 15 = ______

9. True (✓) False (✗):

(a) 4 x 15 = 60 ☐

(b) 1 x 15 = 15 ☐

(c) 6 x 15 = 90 ☐

(d) 7 x 15 = 110 ☐

(e) 2 x 150 = 450 ☐

(f) 5 x 150 = 750 ☐

(g) 6 x 150 = 900 ☐

(h) 60 ÷ 15 = 4 ☐

(i) 120 ÷ 15 = 7 ☐

(j) 45 ÷ 15 = 3 ☐

10. Count by 15s to 150.

www.worldteacherspress.com ©World Teachers Press®

15 x table

0 x 15 = _____	0 x 150 = _____		
1 x 15 = _____	1 x 150 = _____	6 x 15 = _____	6 x 150 = _____
2 x 15 = _____	2 x 150 = _____	7 x 15 = _____	7 x 150 = _____
3 x 15 = _____	3 x 150 = _____	8 x 15 = _____	8 x 150 = _____
4 x 15 = _____	4 x 150 = _____	9 x 15 = _____	9 x 150 = _____
5 x 15 = _____	5 x 150 = _____	10 x 15 = _____	10 x 150 = _____

1. How many days in:

(a) 15 weeks 3 days? __________ *(b) 15 weeks 5 days?* __________

2. Double:

(a) 60 _____ *(b) 45* _____ *(c) 15* _____ *(d) 75* _____

3. Halve:

(a) 120 _____ *(b) 150* _____ *(c) 60* _____ *(d) 30* _____

4. Candies cost 15¢ each. How many can you buy with:

(a) 45¢? _____ *(b) $1.20?* _____ *(c) $1.35?* _____ *(d) $1.50?* _____

5. Each van can carry 15 passengers. How many vans would be needed to carry:

(a) 60 people? _______ *vans* *(b) 150 people?* _______ *vans* *(c) 105 people?* _______ *vans*

6. Complete these:

(a) 48 ÷ 15 = _____ *r* _____ *(b) 107 ÷ 15 =* _____ *r* _____ *(c) 140 ÷ 15 =* _____ *r* _____

7. If you cut these pies into fifteenths, how many slices would you have?

(a) 4 pies _____ *slices* *(b) 5 pies* _____ *slices* *(c) 2 pies* _____ *slices*

(d) 7 pies _____ *slices* *(e) 3 pies* _____ *slices* *(f) 9 pies* _____ *slices*

8. A crate can hold 150 bananas. How many bananas would fit in:

(a) 3 crates? __________ *(b) 6 crates?* __________ *(c) 9 crates?* __________

9. Surf stickers cost $1.50 each. Jordan buys 3 and his friend Jack buys 7 stickers.

(a) How much money does Jordan spend? __________

(b) How much money does Jack spend? __________

(c) How much do Jordan and Jack spend altogether? __________

(d) How much change would Jordan receive from a $10 bill? __________

(e) How much change would Jack receive from a $20 bill? __________

10. Tables Battleship.

4				
3				
2				
1				
	A	B	C	D

4				
3				
2				
1				
	A	B	C	D

4				
3				
2				
1				
	A	B	C	D

4				
3				
2				
1				
	A	B	C	D

©World Teachers Press® www.worldteacherspress.com

15 x table

Set A

1. 30 + 15 = ____
2. 15 x 6 = ____
3. 45 ÷ 15 = ____
4. 30 – 15 = ____
5. 60 + 15 = ____
6. 15 x 3 = ____
7. 105 ÷ 15 = ____
8. 75 – 15 = ____
9. 75 + 15 = ____
10. 15 x 10 = ____
11. 30 ÷ 15 = ____
12. 45 – 15 = ____
13. 45 + 15 = ____
14. 15 x 2 = ____
15. 75 ÷ 15 = ____
16. 15 – 15 = ____
17. 15 + 15 = ____
18. 15 x 4 = ____
19. 135 ÷ 15 = ____
20. 60 – 15 = ____
21. 120 + 15 = ____
22. 15 x 5 = ____
23. 60 ÷ 15 = ____
24. 105 – 15 = ____
25. 90 + 15 = ____
26. 15 x 8 = ____
27. 90 ÷ 15 = ____
28. 90 – 15 = ____
29. 150 x 0 = ____
30. 15 x 9 = ____
31. 15 ÷ 15 = ____
32. 150 – 15 = ____
33. 135 + 15 = ____
34. 15 x 1 = ____
35. 120 ÷ 15 = ____
36. 120 – 15 = ____
37. 105 + 15 = ____
38. 15 x 7 = ____
39. 150 ÷ 15 = ____
40. 135 – 15 = ____

My score:

My time:

____ mins ____ secs

I'm happy | I'm not happy | OOPS! | I didn't understand

The main area for me to work on is:

Set B

1. 120 + 15 = ____
2. 15 x 1 = ____
3. 60 ÷ 15 = ____
4. 60 – 15 = ____
5. 90 + 15 = ____
6. 15 x 7 = ____
7. 90 ÷ 15 = ____
8. 45 – 15 = ____
9. 15 x 0 = ____
10. 15 x 5 = ____
11. 15 ÷ 15 = ____
12. 15 – 15 = ____
13. 135 + 15 = ____
14. 15 x 9 = ____
15. 120 ÷ 15 = ____
16. 30 – 15 = ____
17. 105 + 15 = ____
18. 8 x 15 = ____
19. 150 ÷ 15 = ____
20. 90 – 15 = ____
21. 15 + 15 = ____
22. 15 x 4 = ____
23. 135 ÷ 15 = ____
24. 75 – 15 = ____
25. 30 + 15 = ____
26. 15 x 2 = ____
27. 75 ÷ 15 = ____
28. 120 – 15 = ____
29. 60 + 15 = ____
30. 15 x 10 = ____
31. 30 ÷ 15 = ____
32. 150 – 15 = ____
33. 45 + 15 = ____
34. 15 x 3 = ____
35. 105 ÷ 15 = ____
36. 135 – 15 = ____
37. 75 + 15 = ____
38. 6 x 15 = ____
39. 45 ÷ 15 = ____
40. 105 – 15 = ____

My score:

My time:

____ mins ____ secs

I'm happy | I'm not happy | OOPS! | I didn't understand

The main area for me to work on is:

www.worldteacherspress.com ©World Teachers Press®

Set A

1. Add 20 to the following:

(a) 80 ______ (c) 120 ______

(b) 160 ______ (d) 60 ______

2. How many groups of 20 in:

(a) 60? ______ (c) 100? ______

(b) 160? ______ (d) 120? ______

3. Subtract 20 from the following:

(a) 40 ______ (c) 180 ______

(b) 100 ______ (d) 80 ______

4. Which multiple of 20 is closest to:

(a) 38? ______ (c) 145? ______

(b) 55? ______ (d) 23? ______

5. Add:

(a) 20 + 20 + 20 + 20 = ______

(b) 20 + 20 + 20 + 20 + 20 = ______

(c) 20 + 20 + 20 + 20 + 20 + 20 + 20 + 20 = ______

6. Complete the number sequence:

0, 20, ______, ______, 80, ______, ______, 140, ______, ______, 200

7. Complete the number sentences:

(a) 20 x ______ = 100

(b) 20 x ______ = 180

(c) 20 x ______ = 0

8. Complete the following:

(a) (20 x 7) + 20 = ______

(b) (20 x 4) + 20 = ______

(c) (20 x 10) – 20 = ______

9. True (✓) False (✗):

(a) 3 x 20 = 60 ☐

(b) 9 x 20 = 180 ☐

(c) 6 x 20 = 140 ☐

(d) 8 x 20 = 180 ☐

(e) 7 x 200 = 1400 ☐

(f) 5 x 200 = 1000 ☐

(g) 9 x 200 = 1600 ☐

(h) 100 ÷ 20 = 6 ☐

(i) 180 ÷ 20 = 9 ☐

(j) 140 ÷ 20 = 7 ☐

10. Say your 20 x tables.

Set B

1. Add 20 to the following:

(a) 40 ______ (c) 140 ______

(b) 20 ______ (d) 180 ______

2. How many groups of 20 in:

(a) 80? ______ (c) 200? ______

(b) 180? ______ (d) 40? ______

3. Subtract 20 from the following:

(a) 140 ______ (c) 200 ______

(b) 60 ______ (d) 160 ______

4. Which multiple of 20 is closest to:

(a) 116? ______ (c) 168? ______

(b) 94? ______ (d) 192? ______

5. Add:

(a) 20 + 20 + 20 + 20 + 20 + 20 + 20 = ______

(b) 20 + 20 + 20 = ______

(c) 20 + 20 + 20 + 20 + 20 + 20 + 20 + 20 + 20 = ______

6. Complete the number sequence:

200, ______, ______, 140, ______, 100, ______, ______, ______, 20, ______

7. Complete the number sentences:

(a) 20 x ______ = 120

(b) 20 x ______ = 200

(c) 20 x ______ = 60

8. Complete the following:

(a) (20 x 9) + 20 = ______

(b) (20 x 5) – 20 = ______

(c) (20 x 8) – 20 = ______

9. True (✓) False (✗):

(a) 5 x 20 = 100 ☐

(b) 7 x 20 = 140 ☐

(c) 10 x 20 = 180 ☐

(d) 4 x 200 = 800 ☐

(e) 3 x 200 = 600 ☐

(f) 6 x 200 = 1400 ☐

(g) 8 x 200 = 1600 ☐

(h) 80 ÷ 20 = 4 ☐

(i) 160 ÷ 20 = 8 ☐

(j) 60 ÷ 20 = 3 ☐

10. Count by 20s to 200.

©World Teachers Press® www.worldteacherspress.com

20 x table

0 x 20 = ____	0 x 200 = ____		
1 x 20 = ____	1 x 200 = ____	6 x 20 = ____	6 x 200 = ____
2 x 20 = ____	2 x 200 = ____	7 x 20 = ____	7 x 200 = ____
3 x 20 = ____	3 x 200 = ____	8 x 20 = ____	8 x 200 = ____
4 x 20 = ____	4 x 200 = ____	9 x 20 = ____	9 x 200 = ____
5 x 20 = ____	5 x 200 = ____	10 x 20 = ____	10 x 200 = ____

1. How many days in:

(a) 20 weeks 5 days? ________ *(b) 20 weeks 2 days?* ________

20

2. Double:

(a) 80 ____ *(b) 60* ____ *(c) 20* ____ *(d) 100* ____

3. Halve:

(a) 600 ____ *(b) 160* ____ *(c) 80* ____ *(d) 40* ____

4. Candies cost 20¢ each. How many can you buy with:

(a) $1.00? ____ *(b) 60¢?* ____ *(c) $2.00?* ____ *(d) $1.40?* ____

5. Each bus can carry 20 passengers. How many buses would be needed to carry:

(a) 80 people? ______ *buses* *(b) 140 people?* ______ *buses* *(c) 200 people?* ______ *buses*

6. Complete these:

(a) 68 ÷ 20 = ____ *r* ____ *(b) 105 ÷ 20 =* ____ *r* ____ *(c) 175 ÷ 20 =* ____ *r* ____

7. If you cut these pizzas into twentieths, how many slices would you have?

(a) 2 pizzas ____ *slices* *(c) 5 pizzas* ____ *slices* *(e) 10 pizzas* ____ *slices*

(b) 7 pizzas ____ *slices* *(d) 8 pizzas* ____ *slices* *(f) 4 pizzas* ____ *slices*

8. Each cattle truck can hold 200 cattle. How many cattle would fit in:

(a) 3 trucks? ________ *(b) 7 trucks?* ________ *(c) 10 trucks?* ________

9. Ice creams cost $2.00 each. Sasha buys 6 and her cousin Rosie buys 9 ice creams.

(a) How much money does Sasha spend? ________

(b) How much money does Rosie spend? ________

(c) How much more does Rosie spend than Sasha? ________

(d) How much change would Sasha receive from a $20 bill? ________

(e) How much change would Rosie receive from a $20 bill? ________

10. Tables Battleship.

	A	B	C	D
4				
3				
2				
1				

	A	B	C	D
4				
3				
2				
1				

	A	B	C	D
4				
3				
2				
1				

	A	B	C	D
4				
3				
2				
1				

www.worldteacherspress.com ©World Teachers Press®

Set A

1. 20 + 80 = ____	21. 120 + 20 = ____		
2. 20 x 3 = ____	22. 20 x 2 = ____		
3. 100 ÷ 20 = ____	23. 80 ÷ 20 = ____		
4. 80 – 20 = ____	24. 160 – 20 = ____		
5. 100 + 20 = ____	25. 220 + 20 = ____		
6. 10 x 20 = ____	26. 5 x 20 = ____		
7. 60 ÷ 20 = ____	27. 120 ÷ 20 = ____		
8. 140 – 20 = ____	28. 60 – 20 = ____		
9. 160 + 20 = ____	29. 180 + 20 = ____		
10. 4 x 20 = ____	30. 8 x 20 = ____		
11. 160 ÷ 20 = ____	31. 180 ÷ 20 = ____		
12. 100 – 20 = ____	32. 40 – 20 = ____		
13. 60 + 20 = ____	33. 200 x 0 = ____		
14. 9 x 20 = ____	34. 6 x 20 = ____		
15. 200 ÷ 20 = ____	35. 140 ÷ 20 = ____		
16. 120 – 20 = ____	36. 180 – 20 = ____		
17. 40 + 20 = ____	37. 140 + 20 = ____		
18. 7 x 20 = ____	38. 20 x 0 = ____		
19. 40 ÷ 20 = ____	39. 20 ÷ 20 = ____		
20. 200 – 20 = ____	40. 1 x 200 = ____		

My score: ☐

My time:

____ mins ____ secs

I'm happy / I'm not happy / OOPS! / I didn't understand

The main area for me to work on is:

____

____

____

____

____

____

____

Set B

1. 200 x 1 = ____	21. 140 + 20 = ____
2. 6 x 20 = ____	22. 20 x 3 = ____
3. 140 ÷ 20 = ____	23. 160 ÷ 20 = ____
4. 60 – 20 = ____	24. 100 – 20 = ____
5. 120 + 20 = ____	25. 160 + 20 = ____
6. 4 x 20 = ____	26. 5 x 20 = ____
7. 100 ÷ 20 = ____	27. 60 ÷ 20 = ____
8. 120 – 20 = ____	28. 80 – 20 = ____
9. 180 + 20 = ____	29. 100 + 20 = ____
10. 2 x 20 = ____	30. 8 x 20 = ____
11. 40 – 20 = ____	31. 180 ÷ 20 = ____
12. 240 – 20 = ____	32. 160 – 20 = ____
13. 80 + 20 = ____	33. 40 + 20 = ____
14. 7 x 20 = ____	34. 10 x 20 = ____
15. 80 ÷ 20 = ____	35. 120 ÷ 20 = ____
16. 180 – 20 = ____	36. 140 – 20 = ____
17. 60 + 20 = ____	37. 200 + 20 = ____
18. 9 x 20 = ____	38. 20 x 0 = ____
19. 200 ÷ 20 = ____	39. 20 ÷ 20 = ____
20. 40 – 20 = ____	40. 200 – 20 = ____

My score: ☐

My time:

____ mins ____ secs

I'm happy / I'm not happy / OOPS! / I didn't understand

The main area for me to work on is:

____

____

____

____

____

____

____

©World Teachers Press® www.worldteacherspress.com

Set A

1. Add 25 to the following:

(a) 125 ______ (c) 75 ______

(b) 225 ______ (d) 100 ______

2. How many groups of 25 in:

(a) 100? ______ (c) 250? ______

(b) 75? ______ (d) 175? ______

3. Subtract 25 from the following:

(a) 175 ______ (c) 100 ______

(b) 250 ______ (d) 150 ______

4. Which multiple of 25 is closest to:

(a) 180? ______ (c) 210? ______

(b) 65? ______ (d) 28? ______

20

5. Add:

(a) 25 + 25 + 25 +25 = ______

(b) 25 + 25 + 25 + 25 + 25 = ______

(c) 25 + 25 + 25 + 25 + 25 + 25 + 25 + 25 = ______

6. Complete the number sequence:

0, ______, 50, ______, 100, ______, ______, 175, ______, ______, 250

7. Complete the number sentences:

(a) 25 x ______ = 200

(b) 25 x ______ = 150

(c) 25 x ______ = 225

8. Complete the following:

(a) (25 x 7) + 25 = ______

(b) (25 x 4) + 25 = ______

(c) (25 x 3) – 25 = ______

9. True (✓) False (✗):

(a) 3 x 25 = 75 ☐

(b) 9 x 25 = 200 ☐

(c) 6 x 25 = 150 ☐

(d) 5 x 25 = 125 ☐

(e) 4 x 250 = 1000 ☐

(f) 5 x 250 = 2500 ☐

(g) 9 x 250 = 2250 ☐

(h) 100 ÷ 25 = 4 ☐

(i) 150 ÷ 25 = 7 ☐

(j) 75 ÷ 25 = 3 ☐

10. Say your 25 x tables.

Set B

1. Add 25 to the following:

(a) 150 ______ (c) 50 ______

(b) 175 ______ (d) 200 ______

2. How many groups of 25 in:

(a) 200? ______ (c) 150? ______

(b) 225? ______ (d) 50? ______

3. Subtract 25 from the following:

(a) 50 ______ (c) 200 ______

(b) 75 ______ (d) 125 ______

4. Which multiple of 25 is closest to:

(a) 80? ______ (c) 155? ______

(b) 240? ______ (d) 96? ______

20

5. Add:

(a) 25 + 25 + 25 + 25 + 25 + 25 + 25 = ______

(b) 25 + 25 + 25 + 25 + 25 + 25 = ______

(c) 25 + 25 + 25 + 25 + 25 + 25 + 25 + 25 + 25 = ______

6. Complete the number sequence:

250, ______, ______, 175, ______, 125, ______, ______, ______, 25, ______

7. Complete the number sentences:

(a) 25 x ______ = 100

(b) 25 x ______ = 250

(c) 25 x ______ = 175

8. Complete the following:

(a) (25 x 2) + 25 = ______

(b) (25 x 5) – 25 = ______

(c) (25 x 8) – 25 = ______

9. True (✓) False (✗):

(a) 8 x 25 = 250 ☐

(b) 2 x 25 = 50 ☐

(c) 4 x 25 = 100 ☐

(d) 7 x 250 = 1750 ☐

(e) 5 x 250 = 1500 ☐

(f) 2 x 250 = 500 ☐

(g) 125 ÷ 25 = 6 ☐

(h) 250 ÷ 25 = 10 ☐

(i) 50 ÷ 25 = 2 ☐

(j) 150 ÷ 25 = 6 ☐

10. Count by 25s to 250.

www.worldteacherspress.com ©World Teachers Press®

0 x 25 = ____	*0 x 250 = ____*		
1 x 25 = ____	*1 x 250 = ____*	*6 x 25 = ____*	*6 x 250 = ____*
2 x 25 = ____	*2 x 250 = ____*	*7 x 25 = ____*	*7 x 250 = ____*
3 x 25 = ____	*3 x 250 = ____*	*8 x 25 = ____*	*8 x 250 = ____*
4 x 25 = ____	*4 x 250 = ____*	*9 x 25 = ____*	*9 x 250 = ____*
5 x 25 = ____	*5 x 250 = ____*	*10 x 25 = ____*	*10 x 250 = ____*

1. How many days in:

(a) 25 weeks 3 days? ________ *(b) 25 weeks 5 days? ________*

25

2. Double:

(a) 125 ____ *(b) 75 ____* *(c) 50 ____* *(d) 25 ____*

3. Halve:

(a) 150 ____ *(b) 50 ____* *(c) 200 ____* *(d) 100 ____*

4. Candies cost 25¢ each. How many can you buy with:

(a) $1.25? ____ *(b) $2.25? ____* *(c) $1.00? ____* *(d) $1.50? ____*

5. Each bus can carry 25 passengers. How many buses would be needed to carry:

(a) 250 people? ______ buses *(b) 175 people? ______ buses* *(c) 125 people? ______ buses*

6. Complete these:

(a) 79 ÷ 25 = ____ r ____ *(b) 130 ÷ 25 = ____ r ____* *(c) 215 ÷ 25 = ____ r ____*

7. If you cut these cakes into twenty-fifths, how many slices would you have?

(a) 3 cakes ____ slices *(c) 9 cakes ____ slices* *(e) 5 cakes ____ slices*

(b) 4 cakes ____ slices *(d) 7 cakes ____ slices* *(f) 8 cakes ____ slices*

8. A pallet can hold 250 house bricks. How many bricks would fit on:

(a) 4 pallets? ______ bricks *(b) 6 pallets? ______ bricks* *(c) 10 pallets? ______ bricks*

9. Calendars cost $2.50 each. Jon buys 3 and his friend Sam buys 7 calendars.

(a) How much money does Jon spend? ________

(b) How much money does Sam spend? ________

(c) How much do Jon and Sam spend altogether? ________

(d) How much change would Jon receive from a $10 bill? ________

(e) How much change would Sam receive from a $20 bill? ________

10. Tables Battleship.

	A	B	C	D
4				
3				
2				
1				

	A	B	C	D
4				
3				
2				
1				

	A	B	C	D
4				
3				
2				
1				

	A	B	C	D
4				
3				
2				
1				

©World Teachers Press® www.worldteacherspress.com

25 x table

Set A

1. 50 + 25 = ______
2. 25 x 6 = ______
3. 100 ÷ 25 = ______
4. 250 – 25 = ______
5. 100 + 25 = ______
6. 25 x 10 = ______
7. 200 ÷ 25 = ______
8. 200 – 25 = ______
9. 150 + 25 = ______
10. 25 x 3 = ______
11. 25 ÷ 25 = ______
12. 225 – 25 = ______
13. 25 + 25 = ______
14. 25 x 4 = ______
15. 75 ÷ 25 = ______
16. 175 – 25 = ______
17. 75 + 25 = ______
18. 25 x 7 = ______
19. 150 ÷ 25 = ______
20. 125 – 25 = ______
21. 125 + 25 = ______
22. 25 x 8 = ______
23. 50 ÷ 25 = ______
24. 75 – 25 = ______
25. 175 + 25 = ______
26. 25 x 2 = ______
27. 125 ÷ 25 = ______
28. 25 – 25 = ______
29. 225 + 25 = ______
30. 25 x 5 = ______
31. 175 ÷ 25 = ______
32. 150 – 25 = ______
33. 200 + 25 = ______
34. 25 x 9 = ______
35. 250 ÷ 25 = ______
36. 100 – 25 = ______
37. 2 x 250 = ______
38. 25 x 0 = ______
39. 225 ÷ 25 = ______
40. 50 – 25 = ______

My score:

My time:

_____ mins _____ secs

I'm happy / I'm not happy / OOPS! / I didn't understand

The main area for me to work on is:

____

____

____

____

____

____

____

Set B

1. 250 x 4 = ______
2. 25 x 0 = ______
3. 50 ÷ 25 = ______
4. 100 – 25 = ______
5. 50 + 25 = ______
6. 9 x 25 = ______
7. 125 ÷ 25 = ______
8. 50 – 25 = ______
9. 200 + 25 = ______
10. 25 x 5 = ______
11. 175 ÷ 25 = ______
12. 150 – 25 = ______
13. 225 + 25 = ______
14. 25 x 2 = ______
15. 250 ÷ 25 = ______
16. 25 – 25 = ______
17. 175 + 25 = ______
18. 8 x 25 = ______
19. 225 ÷ 25 = ______
20. 75 – 25 = ______
21. 125 + 25 = ______
22. 7 x 25 = ______
23. 100 ÷ 25 = ______
24. 125 – 25 = ______
25. 75 + 25 = ______
26. 4 x 25 = ______
27. 200 ÷ 25 = ______
28. 175 – 25 = ______
29. 25 + 25 = ______
30. 25 x 3 = ______
31. 25 ÷ 25 = ______
32. 225 – 25 = ______
33. 150 + 25 = ______
34. 25 x 10 = ______
35. 75 ÷ 25 = ______
36. 200 – 25 = ______
37. 100 + 25 = ______
38. 6 x 25 = ______
39. 150 ÷ 25 = ______
40. 250 – 25 = ______

My score:

My time:

_____ mins _____ secs

I'm happy / I'm not happy / OOPS! / I didn't understand

The main area for me to work on is:

____

____

____

____

____

____

____

www.worldteacherspress.com ©World Teachers Press®

Set A

1. Add 50 to the following:

(a) 150 ____ (c) 350 ____

(b) 50 ____ (d) 100 ____

2. How many groups of 50 in:

(a) 200? ____ (c) 350? ____

(b) 100? ____ (d) 400? ____

3. Subtract 50 from the following:

(a) 500 ____ (c) 150 ____

(b) 250 ____ (d) 200 ____

4. Which multiple of 50 is closest to:

(a) 108? ____ (c) 370? ____

(b) 490? ____ (d) 162? ____

5. Add:

(a) 50 + 50 + 50 + 50 = ____

(b) 50 + 50 + 50 + 50 + 50 = ____

(c) 50 + 50 + 50 + 50 + 50 + 50 + 50 + 50 = ____

6. Complete the number sequence:

0, 50, ____, ____, 200, ____, ____, 350, ____, ____, 500

7. Complete the number sentences:

(a) 50 x ____ = 350

(b) 50 x ____ = 250

(c) 50 x ____ = 450

8. Complete the following:

(a) (50 x 6) + 50 = ____

(b) (50 x 9) + 50 = ____

(c) (50 x 3) – 50 = ____

9. True (✓) False (✗):

(a) 3 x 50 = 150 ☐

(b) 9 x 50 = 400 ☐

(c) 6 x 50 = 300 ☐

(d) 8 x 50 = 400 ☐

(e) 7 x 500 = 3500 ☐

(f) 5 x 500 = 2500 ☐

(g) 9 x 500 = 4000 ☐

(h) 100 ÷ 50 = 2 ☐

(i) 350 ÷ 50 = 6 ☐

(j) 50 ÷ 50 = 1 ☐

10. Say your 50 x tables.

Set B

1. Add 50 to the following:

(a) 450 ____ (c) 300 ____

(b) 250 ____ (d) 200 ____

2. How many groups of 50 in:

(a) 300? ____ (c) 450? ____

(b) 500? ____ (d) 250? ____

3. Subtract 50 from the following:

(a) 450 ____ (c) 100 ____

(b) 300 ____ (d) 550 ____

4. Which multiple of 50 is closest to:

(a) 415? ____ (c) 142? ____

(b) 280? ____ (d) 519? ____

5. Add:

(a) 50 + 50 + 50 + 50 + 50 + 50 + 50 = ____

(b) 50 + 50 + 50 + 50 + 50 + 50 = ____

(c) 50 + 50 + 50 + 50 + 50 + 50 + 50 + 50 + 50 = ____

6. Complete the number sequence:

____, 450, ____, ____, 300, ____, ____, ____, 100, ____, ____

7. Complete the number sentences:

(a) 50 x ____ = 500

(b) 50 x ____ = 200

(c) 50 x ____ = 400

8. Complete the following:

(a) (50 x 4) + 50 = ____

(b) (50 x 10) – 50 = ____

(c) (50 x 7) – 50 = ____

9. True (✓) False (✗):

(a) 10 x 50 = 500 ☐

(b) 7 x 50 = 350 ☐

(c) 5 x 50 = 200 ☐

(d) 2 x 500 = 1500 ☐

(e) 4 x 500 = 2000 ☐

(f) 6 x 500 = 3500 ☐

(g) 3 x 500 = 1500 ☐

(h) 250 ÷ 50 = 4 ☐

(i) 150 ÷ 50 = 3 ☐

(j) 400 ÷ 50 = 8 ☐

10. Count by 50s to 500.

©World Teachers Press® www.worldteacherspress.com

50 x table

0 x 50 = _____	0 x 500 = _____		
1 x 50 = _____	1 x 500 = _____	6 x 50 = _____	6 x 500 = _____
2 x 50 = _____	2 x 500 = _____	7 x 50 = _____	7 x 500 = _____
3 x 50 = _____	3 x 500 = _____	8 x 50 = _____	8 x 500 = _____
4 x 50 = _____	4 x 500 = _____	9 x 50 = _____	9 x 500 = _____
5 x 50 = _____	5 x 500 = _____	10 x 50 = _____	10 x 500 = _____

1. How many days in:

(a) 50 weeks 4 days? ____________ (b) 50 weeks 1 day? ____________

50

2. Double:

(a) 150 _____ (b) 200 _____ (c) 50 _____ (d) 400 _____

3. Halve:

(a) 500 _____ (b) 300 _____ (c) 100 _____ (d) 200 _____

4. Pencils cost 50¢ each. How many can you buy with:

(a) $2.50? _____ (b) $4.50? _____ (c) $1.50? _____ (d) $3.00? _____

5. Each bus can carry 50 passengers. How many buses would be needed to carry:

(a) 150 people? ________ buses (b) 300 people? ________ buses (c) 450 people? ________ buses

6. Complete these:

(a) 120 ÷ 50 = _____ r _____ (b) 380 ÷ 50 = _____ r _____ (c) 460 ÷ 50 = _____ r _____

7. If you cut these yard-long gummy snakes into fiftieths, how many pieces would you have?

(a) 3 snakes _____ pieces (b) 8 snakes _____ pieces (c) 10 snakes _____ pieces

(d) 5 snakes _____ pieces (e) 7 snakes _____ pieces (f) 9 snakes _____ pieces

8. Each submarine can hold 500 sailors. How many sailors would fit in:

(a) 6 submarines? ________ sailors (b) 10 submarines? ________ sailors (c) 7 submarines? ________ sailors

9. Magazines cost $5.00 each. Kate buys 5 and Harley buys 9 magazines.

(a) How much money does Kate spend? ____________

(b) How much money does Harley spend? ____________

(c) How much more does Harley spend than Kate? ____________

(d) How much change would Kate receive from a $50 bill? ____________

(e) How much change would Harley receive from a $50 bill? ____________

10. Tables Battleship.

	A	B	C	D
4				
3				
2				
1				

	A	B	C	D
4				
3				
2				
1				

	A	B	C	D
4				
3				
2				
1				

	A	B	C	D
4				
3				
2				
1				

www.worldteacherspress.com ©World Teachers Press®

Set A

1. 100 + 50 = ____	21. 250 + 50 = ____
2. 50 x 6 = ____	22. 50 x 8 = ____
3. 200 ÷ 50 = ____	23. 100 ÷ 50 = ____
4. 500 – 50 = ____	24. 150 – 50 = ____
5. 200 + 50 = ____	25. 350 + 50 = ____
6. 50 x 10 = ____	26. 500 x 0 = ____
7. 400 ÷ 50 = ____	27. 250 ÷ 50 = ____
8. 400 – 50 = ____	28. 50 – 50 = ____
9. 300 + 50 = ____	29. 450 + 50 = ____
10. 50 x 3 = ____	30. 50 x 5 = ____
11. 50 ÷ 50 = ____	31. 350 ÷ 50 = ____
12. 450 – 50 = ____	32. 300 – 50 = ____
13. 50 + 50 = ____	33. 400 + 50 = ____
14. 50 x 4 = ____	34. 50 x 9 = ____
15. 150 ÷ 50 = ____	35. 500 ÷ 50 = ____
16. 350 – 50 = ____	36. 200 – 50 = ____
17. 150 + 50 = ____	37. 50 x 0 = ____
18. 50 x 7 = ____	38. 50 x 8 = ____
19. 300 ÷ 50 = ____	39. 450 ÷ 50 = ____
20. 250 – 50 = ____	40. 100 – 50 = ____

My score: ☐

My time:

____ mins ____ secs

I'm happy · I'm not happy · OOPS! · I didn't understand

The main area for me to work on is:

____

____

____

____

____

____

____

Set B

1. 500 x 1 = ____	21. 4 x 500 = ____
2. 50 x 0 = ____	22. 50 x 7 = ____
3. 100 ÷ 50 = ____	23. 200 ÷ 50 = ____
4. 200 – 50 = ____	24. 250 – 50 = ____
5. 100 + 50 = ____	25. 150 + 50 = ____
6. 50 x 9 = ____	26. 50 x 4 = ____
7. 250 ÷ 50 = ____	27. 400 ÷ 50 = ____
8. 100 – 50 = ____	28. 350 – 50 = ____
9. 400 + 50 = ____	29. 50 + 50 = ____
10. 50 x 5 = ____	30. 50 x 5 = ____
11. 350 ÷ 50 = ____	31. 50 ÷ 50 = ____
12. 300 – 50 = ____	32. 450 – 50 = ____
13. 450 + 50 = ____	33. 300 + 50 = ____
14. 50 x 2 = ____	34. 50 x 10 = ____
15. 500 ÷ 50 = ____	35. 150 ÷ 50 = ____
16. 50 – 50 = ____	36. 400 – 50 = ____
17. 350 + 50 = ____	37. 200 + 50 = ____
18. 50 x 8 = ____	38. 50 x 6 = ____
19. 450 ÷ 50 = ____	39. 300 ÷ 50 = ____
20. 150 – 50 = ____	40. 500 – 50 = ____

My score: ☐

My time:

____ mins ____ secs

I'm happy · I'm not happy · OOPS! · I didn't understand

The main area for me to work on is:

____

____

____

____

____

____

____

©World Teachers Press® www.worldteacherspress.com

2 and 3 times tables

Set A

1. 2 x 7 = ____
2. 3 x 0 = ____
3. 16 ÷ 2= ____
4. 21 ÷ 3= ____
5. 2 + 2 + 2 = ____
6. 3 + 3 + 3 + 3 = ____
7. 10 – 2 = ____
8. 15 – 3 = ____
9. 6 x 20 = ____
10. 4 x 30 = ____
11. Double 20 = ____
12. Double 30 = ____
13. 2 x ____ = 12
14. 3 x ____ = 18
15. (2 x 5) + 2 = ____
16. (3 x 9) + 3 = ____
17. (2 x 10) + 2 = ____
18. (3 x 8) + 3 = ____
19. (2 x 7) – 2 = ____
20. (3 x 6) – 3 = ____
21. (2 x 9) – 2 = ____
22. (3 x 4) – 3 = ____
23. 9 ÷ 2 = ____ r ____
24. 17 ÷ 3 = ____ r ____
25. 2 x 40 = ____
26. 3 x 60 = ____
27. 2 x ____ = 16
28. 3 x ____ = 21
29. 2 x ____ = 20
30. 3 x ____ = 0
31. Half 10 = ____
32. Half 6 = ____
33. 18 ÷ 2 = ____
34. 12 ÷ 3 = ____
35. 14 – 2 = ____
36. 18 – 3 = ____
37. 20 ÷ 2 = ____
38. 30 ÷ 3 = ____
39. 2 x 20 = ____
40. 3 x 30 = ____

My score

My time: ____ mins ____ secs

I'm happy / I'm not happy / OOPS! / I didn't understand

The main area for me to work on is: ______________________

Set B

1. 2 x 4 = ____
2. 3 x 9 = ____
3. 18 ÷ 2 = ____
4. 27 ÷ 3 = ____
5. 2 + 2 + 2 + 2 = ____
6. 3 + 3 + 3 = ____
7. 18 – 2 = ____
8. 12 – 3 = ____
9. 4 x 20 = ____
10. 7 x 30 = ____
11. Double 40 = ____
12. Double 60 = ____
13. 2 x ____ = 16
14. 3 x ____ = 24
15. (2 x 7) + 2= ____
16. (3 x 4) + 3 = ____
17. (2 x 9) + 2 = ____
18. (3 x 6) + 3= ____
19. (2 x 5) – 2= ____
20. (3 x 8) – 3= ____
21. (2 x 10) – 2= ____
22. (3 x 10) – 3= ____
23. 7 ÷ 2 = ____
24. 22 ÷ 3 = ____
25. 2 x 50 = ____
26. 3 x 70 = ____
27. 2 x ____ = 20
28. 3 x ____ = 24
29. 2 x ____ = 18
30. 3 x ____ = 27
31. Half 8 = ____
32. Half 30 = ____
33. 20 ÷ 2 = ____
34. 18 ÷ 3 = ____
35. 10 – 2 = ____
36. 24 – 3 = ____
37. 14 ÷ 2 = ____
38. 3 ÷ 3 = ____
39. 7 x 20 = ____
40. 6 x 30 = ____

My score

My time: ____ mins ____ secs

I'm happy / I'm not happy / OOPS! / I didn't understand

The main area for me to work on is: ______________________

www.worldteacherspress.com ©World Teachers Press®

4 and 5 times tables

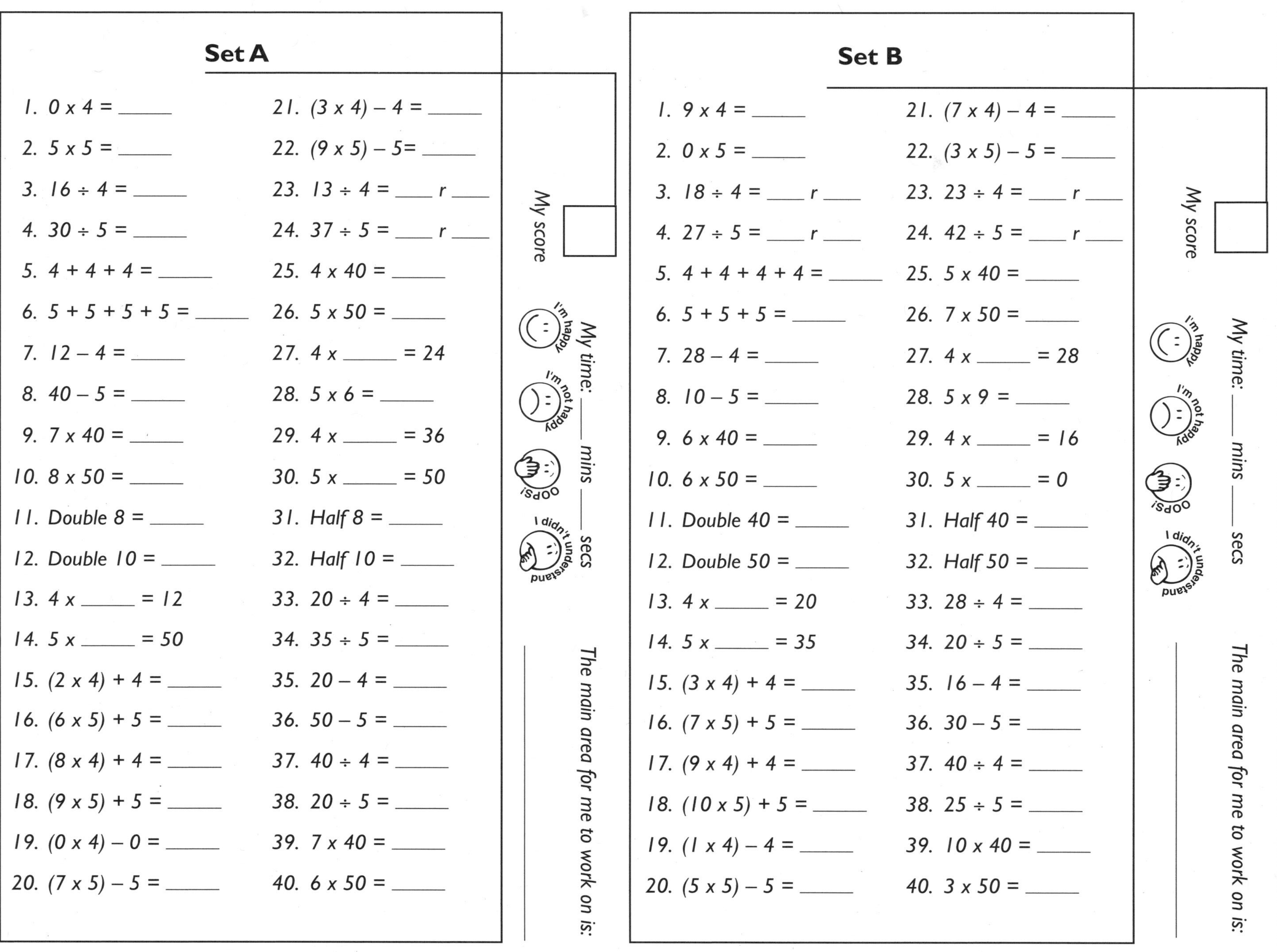

©World Teachers Press® www.worldteacherspress.com

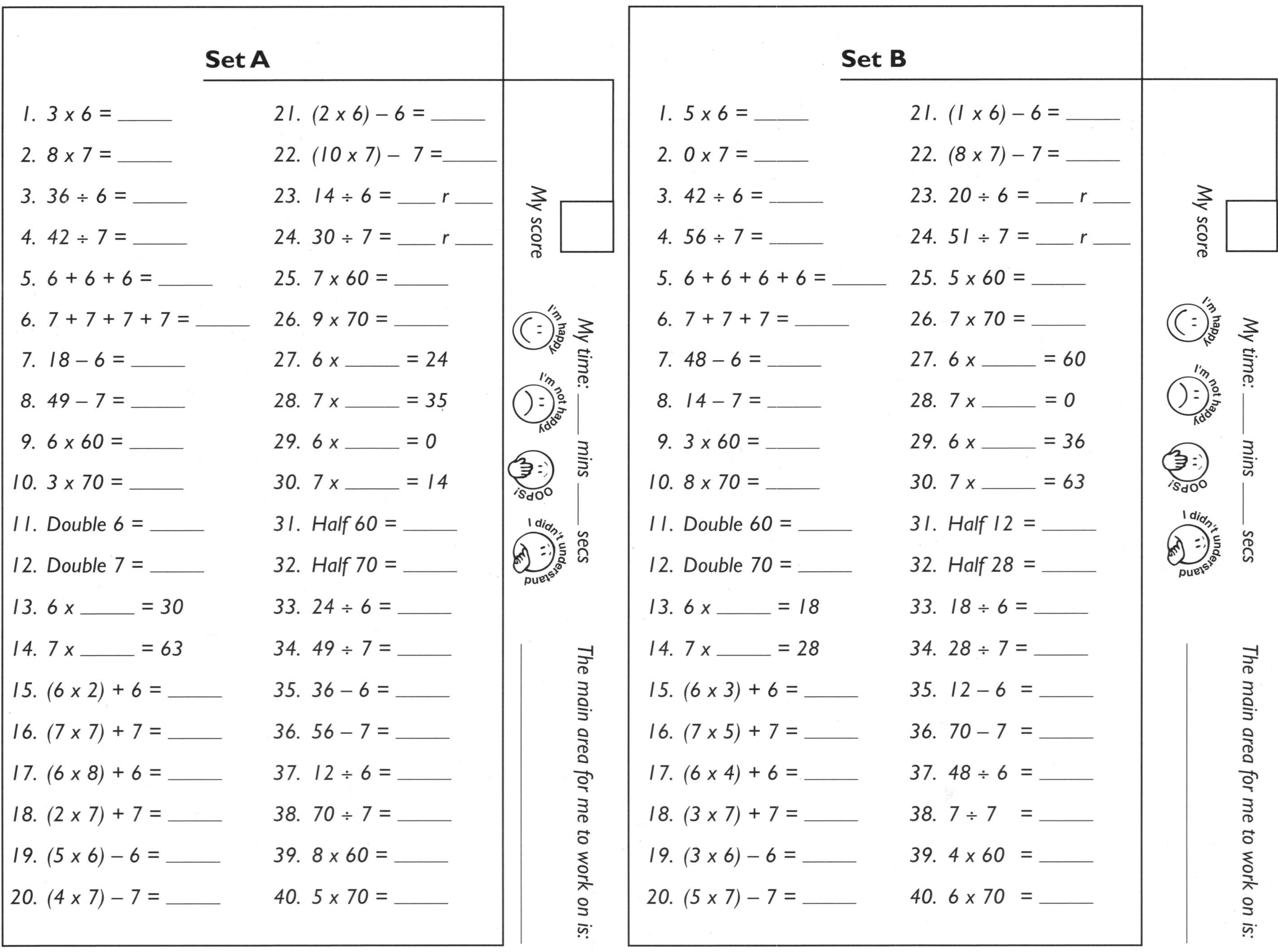

Set A

My score

My time: ____ mins ____ secs

I'm happy · I'm not happy · OOPS! · I didn't understand

The main area for me to work on is:

1. 3 x 6 = ____
2. 8 x 7 = ____
3. 36 ÷ 6 = ____
4. 42 ÷ 7 = ____
5. 6 + 6 + 6 = ____
6. 7 + 7 + 7 + 7 = ____
7. 18 – 6 = ____
8. 49 – 7 = ____
9. 6 x 60 = ____
10. 3 x 70 = ____
11. Double 6 = ____
12. Double 7 = ____
13. 6 x ____ = 30
14. 7 x ____ = 63
15. (6 x 2) + 6 = ____
16. (7 x 7) + 7 = ____
17. (6 x 8) + 6 = ____
18. (2 x 7) + 7 = ____
19. (5 x 6) – 6 = ____
20. (4 x 7) – 7 = ____
21. (2 x 6) – 6 = ____
22. (10 x 7) – 7 = ____
23. 14 ÷ 6 = ____ r ____
24. 30 ÷ 7 = ____ r ____
25. 7 x 60 = ____
26. 9 x 70 = ____
27. 6 x ____ = 24
28. 7 x ____ = 35
29. 6 x ____ = 0
30. 7 x ____ = 14
31. Half 60 = ____
32. Half 70 = ____
33. 24 ÷ 6 = ____
34. 49 ÷ 7 = ____
35. 36 – 6 = ____
36. 56 – 7 = ____
37. 12 ÷ 6 = ____
38. 70 ÷ 7 = ____
39. 8 x 60 = ____
40. 5 x 70 = ____

Set B

My score

My time: ____ mins ____ secs

I'm happy · I'm not happy · OOPS! · I didn't understand

The main area for me to work on is:

1. 5 x 6 = ____
2. 0 x 7 = ____
3. 42 ÷ 6 = ____
4. 56 ÷ 7 = ____
5. 6 + 6 + 6 + 6 = ____
6. 7 + 7 + 7 = ____
7. 48 – 6 = ____
8. 14 – 7 = ____
9. 3 x 60 = ____
10. 8 x 70 = ____
11. Double 60 = ____
12. Double 70 = ____
13. 6 x ____ = 18
14. 7 x ____ = 28
15. (6 x 3) + 6 = ____
16. (7 x 5) + 7 = ____
17. (6 x 4) + 6 = ____
18. (3 x 7) + 7 = ____
19. (3 x 6) – 6 = ____
20. (5 x 7) – 7 = ____
21. (1 x 6) – 6 = ____
22. (8 x 7) – 7 = ____
23. 20 ÷ 6 = ____ r ____
24. 51 ÷ 7 = ____ r ____
25. 5 x 60 = ____
26. 7 x 70 = ____
27. 6 x ____ = 60
28. 7 x ____ = 0
29. 6 x ____ = 36
30. 7 x ____ = 63
31. Half 12 = ____
32. Half 28 = ____
33. 18 ÷ 6 = ____
34. 28 ÷ 7 = ____
35. 12 – 6 = ____
36. 70 – 7 = ____
37. 48 ÷ 6 = ____
38. 7 ÷ 7 = ____
39. 4 x 60 = ____
40. 6 x 70 = ____

www.worldteacherspress.com ©World Teachers Press®

8 and 9 times tables

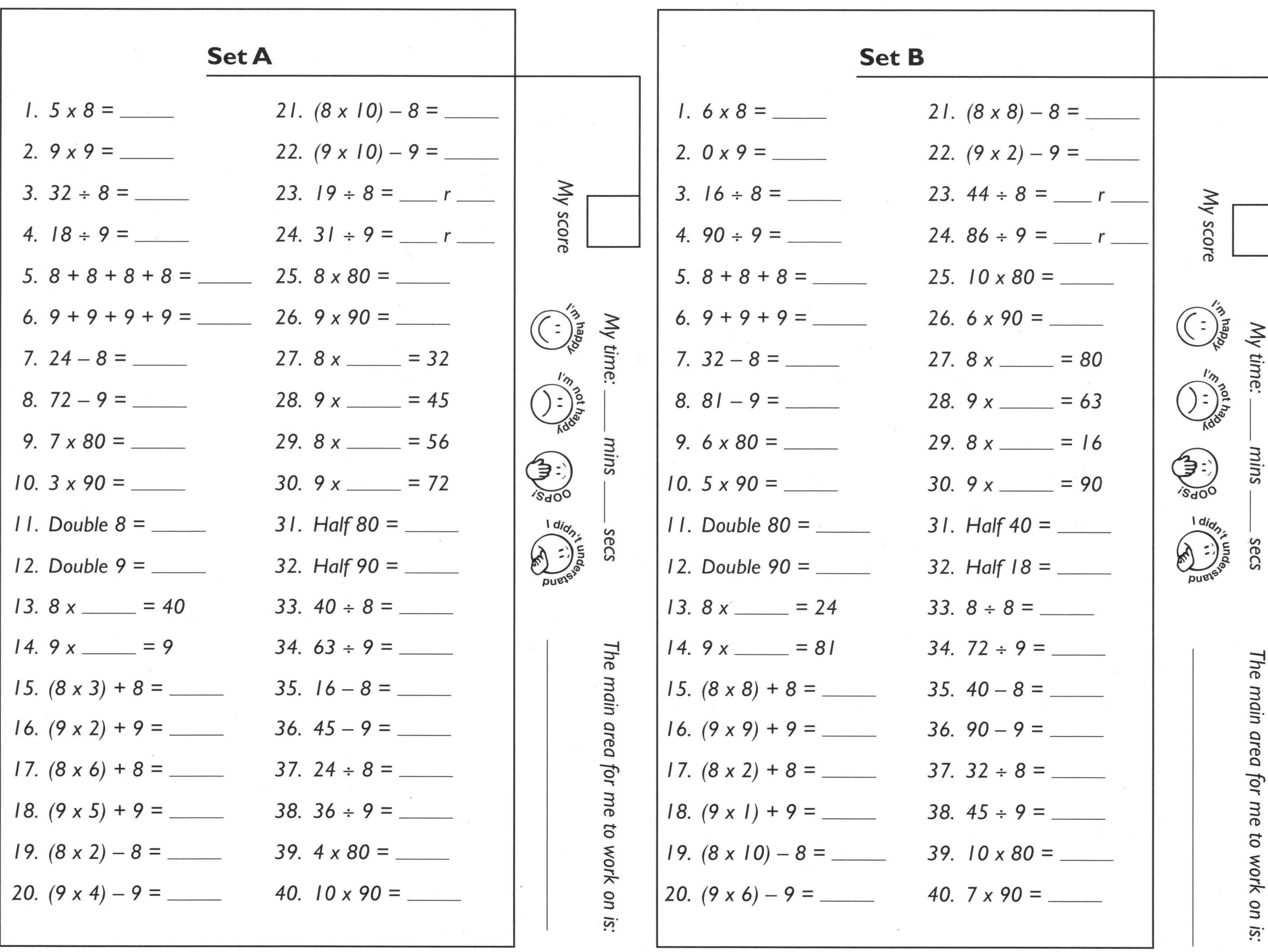

Set A

My score

My time: ____ mins ____ secs

The main area for me to work on is: ______________

1. 5 x 8 = ______
2. 9 x 9 = ______
3. 32 ÷ 8 = ______
4. 18 ÷ 9 = ______
5. 8 + 8 + 8 + 8 = ______
6. 9 + 9 + 9 + 9 = ______
7. 24 – 8 = ______
8. 72 – 9 = ______
9. 7 x 80 = ______
10. 3 x 90 = ______
11. Double 8 = ______
12. Double 9 = ______
13. 8 x ______ = 40
14. 9 x ______ = 9
15. (8 x 3) + 8 = ______
16. (9 x 2) + 9 = ______
17. (8 x 6) + 8 = ______
18. (9 x 5) + 9 = ______
19. (8 x 2) – 8 = ______
20. (9 x 4) – 9 = ______
21. (8 x 10) – 8 = ______
22. (9 x 10) – 9 = ______
23. 19 ÷ 8 = ____ r ____
24. 31 ÷ 9 = ____ r ____
25. 8 x 80 = ______
26. 9 x 90 = ______
27. 8 x ______ = 32
28. 9 x ______ = 45
29. 8 x ______ = 56
30. 9 x ______ = 72
31. Half 80 = ______
32. Half 90 = ______
33. 40 ÷ 8 = ______
34. 63 ÷ 9 = ______
35. 16 – 8 = ______
36. 45 – 9 = ______
37. 24 ÷ 8 = ______
38. 36 ÷ 9 = ______
39. 4 x 80 = ______
40. 10 x 90 = ______

Set B

My score

My time: ____ mins ____ secs

The main area for me to work on is: ______________

1. 6 x 8 = ______
2. 0 x 9 = ______
3. 16 ÷ 8 = ______
4. 90 ÷ 9 = ______
5. 8 + 8 + 8 = ______
6. 9 + 9 + 9 = ______
7. 32 – 8 = ______
8. 81 – 9 = ______
9. 6 x 80 = ______
10. 5 x 90 = ______
11. Double 80 = ______
12. Double 90 = ______
13. 8 x ______ = 24
14. 9 x ______ = 81
15. (8 x 8) + 8 = ______
16. (9 x 9) + 9 = ______
17. (8 x 2) + 8 = ______
18. (9 x 1) + 9 = ______
19. (8 x 10) – 8 = ______
20. (9 x 6) – 9 = ______
21. (8 x 8) – 8 = ______
22. (9 x 2) – 9 = ______
23. 44 ÷ 8 = ____ r ____
24. 86 ÷ 9 = ____ r ____
25. 10 x 80 = ______
26. 6 x 90 = ______
27. 8 x ______ = 80
28. 9 x ______ = 63
29. 8 x ______ = 16
30. 9 x ______ = 90
31. Half 40 = ______
32. Half 18 = ______
33. 8 ÷ 8 = ______
34. 72 ÷ 9 = ______
35. 40 – 8 = ______
36. 90 – 9 = ______
37. 32 ÷ 8 = ______
38. 45 ÷ 9 = ______
39. 10 x 80 = ______
40. 7 x 90 = ______

©World Teachers Press® www.worldteacherspress.com

10 and 12 times tables

Set A

My score

My time: ____ mins ____ secs

I'm happy / I'm not happy / OOPS! / I didn't understand

The main area for me to work on is: ____________________

1. 7 x 10 = ____
2. 0 x 12 = ____
3. 80 ÷ 10 = ____
4. 72 ÷ 12 = ____
5. 10 + 10 + 10 = ____
6. 12 + 12 + 12 = ____
7. 60 – 10 = ____
8. 24 – 12 = ____
9. 4 x 100 = ____
10. 2 x 120 = ____
11. Double 10 = ____
12. Double 12 = ____
13. 10 x ____ = 30
14. 12 x ____ = 108
15. (10 x 2) + 10 = ____
16. (12 x 3) + 12 = ____
17. (10 x 5) + 10 = ____
18. (12 x 2) + 12 = ____
19. (10 x 7) – 10 = ____
20. (12 x 4) – 12 = ____
21. (10 x 2) – 10 = ____
22. (12 x 6) – 12 = ____
23. 87 ÷ 10 = ____ r ____
24. 39 ÷ 12 = ____ r ____
25. 10 x 100 = ____
26. 3 x 120 = ____
27. 10 x ____ = 50
28. 12 x ____ = 48
29. 10 x ____ = 0
30. 12 x ____ = 96
31. Half 10 = ____
32. Half 12 = ____
33. 90 ÷ 10 = ____
34. 24 ÷ 12 = ____
35. 100 – 10 = ____
36. 60 – 12 = ____
37. 40 ÷ 10 = ____
38. 120 ÷ 12 = ____
39. 6 x 100 = ____
40. 10 x 120 = ____

Set B

My score

My time: ____ mins ____ secs

I'm happy / I'm not happy / OOPS! / I didn't understand

The main area for me to work on is: ____________________

1. 0 x 10 = ____
2. 8 x 12 = ____
3. 100 ÷ 10 = ____
4. 84 ÷ 12 = ____
5. 10 + 10 – 10 = ____
6. 12 + 12 + 12 = ____
7. 20 – 10 = ____
8. 36 – 12 = ____
9. 5 x 100 = ____
10. 0 x 120 = ____
11. Double 100 = ____
12. Double 120 = ____
13. 10 x ____ = 60
14. 12 x ____ = 60
15. (10 x 4) + 10 = ____
16. (12 x 6) + 12 = ____
17. (10 x 8) + 10 = ____
18. (12 x 5) + 12 = ____
19. (10 x 10) – 10 = ____
20. (12 x 1) – 12 = ____
21. (10 x 9) – 10 = ____
22. (12 x 5) – 12 = ____
23. 94 ÷ 10 = ____ r ____
24. 50 ÷ 12 = ____ r ____
25. 2 x 100 = ____
26. 5 x 120 = ____
27. 10 x ____ = 70
28. 12 x ____ = 0
29. 10 x ____ = 100
30. 12 x ____ = 120
31. Half 100 = ____
32. Half 120 = ____
33. 40 ÷ 10 = ____
34. 48 ÷ 12 = ____
35. 70 – 10 = ____
36. 72 – 12 = ____
37. 60 ÷ 10 = ____
38. 48 ÷ 12 = ____
39. 5 x 100 = ____
40. 4 x 120 = ____

www.worldteacherspress.com ©World Teachers Press®

.... page 16

Set A

1. (a) 16 (c) 22
   (b) 6 (d) 20
2. (a) 3 (c) 6
   (b) 10 (d) 8
3. (a) 16 (c) 10
   (b) 6 (d) 8
4. (a) 4, 6 (c) 10, 12
   (b) 2, 4 (d) 20, 22
5. (a) 8 (c) 16
   (b) 10
6. 6, 10, 12, 16, 18, 22, 24
7. (a) 7 (c) 2
   (b) 5
8. (a) 12 (c) 20
   (b) 12
9. (a) ✔ (f) ✘
   (b) ✘ (g) ✔
   (c) ✘ (h) ✔
   (d) ✔ (i) ✔
   (e) ✔ (j) ✘
10. Teacher check

Set B

1. (a) 12 (c) 20
   (b) 10 (d) 22
2. (a) 8 (c) 5
   (b) 1 (d) 7
3. (a) 2 (c) 18
   (b) 12 (d) 20
4. (a) 8, 10 (c) 6, 8
   (b) 12, 14 (d) 14, 16
5. (a) 14 (c) 18
   (b) 12
6. 18, 16, 12, 8, 6, 4, 0
7. (a) 4 (c) 9
   (b) 2
8. (a) 10 (c) 6
   (b) 14
9. (a) ✔ (f) ✘
   (b) ✔ (g) ✔
   (c) ✘ (h) ✔
   (d) ✔ (i) ✔
   (e) ✔ (j) ✘
10. Teacher check

.... page 17

1. (a) 18 (b) 20
2. (a) 80 (c) 120
   (b) 200 (d) 40
3. (a) 80 (c) 40
   (b) 60 (d) 90
4. (a) 11 (c) 9
   (b) 5
5. (a) 5 (c) 7
   (b) 9
6. (a) 8 r1 (c) 10 r1
   (b) 6 r1
7. (a) 14 (d) 8
   (b) 18 (e) 10
   (c) 6 (f) 4
8. (a) 120 (c) 160
   (b) 60
9. (a) $1.40 (d) $3.60
   (b) $1.80 (e) $3.20
   (c) $3.20
10. Teacher check

.... page 18

Set A

1. 8	21. 6
2. 14	22. 16
3. 6	23. 1
4. 14	24. 22
5. 12	25. 10
6. 0	26. 18
7. 4	27. 9
8. 18	28. 10
9. 16	29. 14
10. 10	30. 12
11. 12	31. 3
12. 12	32. 6
13. 20	33. 18
14. 20	34. 0
15. 5	35. 10
16. 16	36. 8
17. 24	37. 22
18. 8	38. 20
19. 8	39. 7
20. 20	40. 4

Set B

1. 18	21. 14
2. 8	22. 18
3. 10	23. 6
4. 10	24. 12
5. 22	25. 10
6. 16	26. 14
7. 5	27. 8
8. 22	28. 16
9. 20	29. 12
10. 20	30. 12
11. 7	31. 1
12. 4	32. 18
13. 24	33. 8
14. 4	34. 40
15. 12	35. 9
16. 6	36. 20
17. 16	37. 4
18. 10	38. 0
19. 3	39. 4
20. 8	40. 2

.... page 19

Set A

1. (a) 15 (c) 24
   (b) 30 (d) 9
2. (a) 10 (c) 3
   (b) 5 (d) 6
3. (a) 24 (c) 18
   (b) 12 (d) 9
4. (a) 18 (c) 24
   (b) 15 (d) 30
5. (a) 12 (c) 24
   (b) 15
6. 9, 15, 18, 24, 27
7. (a) 9 (c) 3
   (b) 3
8. (a) 18 (c) 24
   (b) 15
9. (a) ✔ (f) ✘
   (b) ✔ (g) ✔
   (c) ✘ (h) ✘
   (d) ✔ (i) ✔
   (e) ✘ (j) ✔
10. Teacher check

Set B

1. (a) 12 (c) 6
   (b) 21 (d) 18
2. (a) 7 (c) 4
   (b) 8 (d) 1
3. (a) 6 (c) 27
   (b) 15 (d) 3
4. (a) 9 (c) 30
   (b) 12 (d) 18
5. (a) 9 (c) 27
   (b) 18
6. 27, 24, 18, 12, 9, 6, 0
7. (a) 4 (c) 5
   (b) 3
8. (a) 12 (c) 27
   (b) 24
9. (a) ✘ (f) ✘
   (b) ✔ (g) ✔
   (c) ✔ (h) ✔
   (d) ✔ (i) ✔
   (e) ✔ (j) ✘
10. Teacher check

.... page 20

1. (a) 26 (b) 24
2. (a) 120 (c) 240
   (b) 180 (d) 30
3. (a) 150 (c) 120
   (b) 90 (d) 210
4. (a) 5 (c) 9
   (b) 10
5. (a) 5 (c) 4
   (b) 9
6. (a) 9 r2 (c) 6 r1
   (b) 3 r2
7. (a) 12 (d) 6
   (b) 18 (e) 15
   (c) 27 (f) 30
8. (a) 180 (c) 210
   (b) 120
9. (a) $2.40 (d) $2.60
   (b) $1.50 (e) $3.50
   (c) 90¢
10. Teacher check

.... page 21

Set A

1. 9	21. 6
2. 21	22. 15
3. 10	23. 9
4. 15	24. 18
5. 12	25. 24
6. 24	26. 12
7. 4	27. 5
8. 24	28. 12
9. 15	29. 30
10. 27	30. 6
11. 7	31. 2
12. 21	32. 6
13. 18	33. 27
14. 0	34. 9
15. 6	35. 3
16. 27	36. 0
17. 21	37. 33
18. 18	38. 0
19. 8	39. 1
20. 9	40. 3

Set B

1. 12	21. 21
2. 18	22. 21
3. 5	23. 7
4. 24	24. 30
5. 24	25. 27
6. 0	26. 30
7. 8	27. 2
8. 15	28. 18
9. 30	29. 33
10. 24	30. 90
11. 3	31. 6
12. 27	32. 21
13. 9	33. 6
14. 6	34. 27
15. 3	35. 4
16. 6	36. 33
17. 15	37. 18
18. 12	38. 9
19. 1	39. 9
20. 9	40. 12

.... page 22

Set A

1. (a) 20 (c) 36
   (b) 12 (d) 32
2. (a) 6 (c) 4
   (b) 8 (d) 2
3. (a) 24 (c) 8
   (b) 32 (d) 16
4. (a) 40 (c) 24
   (b) 16 (d) 32
5. (a) 12 (c) 32
   (b) 20
6. 8, 16, 24, 28, 36
7. (a) 5 (c) 3
   (b) 0
8. (a) 32 (c) 36
   (b) 24
9. (a) ✔ (f) ✔
   (b) ✘ (g) ✘
   (c) ✔ (h) ✔
   (d) ✘ (i) ✔
   (e) ✔ (j) ✘
10. Teacher check

Set B

1. (a) 24 (c) 40
   (b) 28 (d) 44
2. (a) 10 (c) 3
   (b) 5 (d) 9
3. (a) 12 (c) 4
   (b) 28 (d) 36
4. (a) 12 (c) 20
   (b) 4 (d) 28
5. (a) 28 (c) 40
   (b) 24
6. 36, 32, 24, 20, 12, 8, 0
7. (a) 9 (c) 7
   (b) 4
8. (a) 20 (c) 20
   (b) 36
9. (a) ✘ (f) ✔
   (b) ✘ (g) ✘
   (c) ✔ (h) ✘
   (d) ✔ (i) ✔
   (e) ✔ (j) ✔
10. Teacher check

.... page 23

1. (a) 31 (b) 34
2. (a) 160 (c) 80
   (b) 240 (d) 320
3. (a) 200 (c) 40
   (b) 160 (d) 140
4. (a) 5 (c) 9
   (b) 10
5. (a) 4 (c) 7
   (b) 9
6. (a) 6 r1 (c) 4 r2
   (b) 7 r2
7. (a) 12 (d) 16
   (b) 36 (e) 20
   (c) 28 (f) 32
8. (a) 200 (c) 320
   (b) 120
9. (a) $3.60 (d) $6.40
   (b) $2.40 (e) $7.60
   (c) $6.00
10. Teacher check

.... page 24

Set A

1. 8	21. 20
2. 20	22. 32
3. 10	23. 5
4. 8	24. 24
5. 16	25. 36
6. 28	26. 16
7. 4	27. 7
8. 20	28. 12
9. 32	29. 44
10. 12	30. 24
11. 2	31. 9
12. 4	32. 16
13. 12	33. 28
14. 36	34. 8
15. 3	35. 6
16. 32	36. 0
17. 24	37. 40
18. 40	38. 0
19. 8	39. 10
20. 36	40. 28

Set B

1. 16	21. 36
2. 28	22. 320
3. 6	23. 7
4. 4	24. 12
5. 24	25. 44
6. 36	26. 32
7. 2	27. 5
8. 28	28. 16
9. 20	29. 44
10. 12	30. 200
11. 8	31. 9
12. 40	32. 8
13. 12	33. 28
14. 24	34. 0
15. 3	35. 1
16. 32	36. 24
17. 8	37. 40
18. 40	38. 16
19. 4	39. 10
20. 20	40. 36

.... page 25

Set A

1. (a) 20 (c) 40
   (b) 25 (d) 35
2. (a) 2 (c) 9

©World Teachers Press® www.worldteacherspress.com

Answers

(b) 5 (d) 6
3. (a) 25 (c) 45
(b) 20 (d) 10
4. (a) 45 (c) 15
(b) 35 (d) 40
5. (a) 15 (c) 40
(b) 30
6. 10, 20, 25, 35, 45, 50
7. (a) 10 (c) 6
(b) 0
8. (a) 35 (c) 35
(b) 20
9. (a) ✔ (f) ✔
(b) ✔ (g) ✔
(c) ✘ (h) ✔
(d) ✔ (i) ✘
(e) ✘ (j) ✔
10. Teacher check

Set B

1. (a) 10 (c) 45
(b) 30 (d) 15
2. (a) 10 (c) 4
(b) 7 (d) 8
3. (a) 5 (c) 15
(b) 40 (d) 30
4. (a) 50 (c) 10
(b) 25 (d) 15
5. (a) 35 (c) 45
(b) 20
6. 50, 40, 30, 25, 10, 5
7. (a) 9 (c) 5
(b) 2
8. (a) 50 (c) 30
(b) 45
9. (a) ✘ (f) ✘
(b) ✘ (g) ✔
(c) ✔ (h) ✔
(d) ✘ (i) ✘
(e) ✔ (j) ✔
10. Teacher check

.... page 26

1. (a) 40 (b) 37
2. (a) 300 (c) 50
(b) 500 (d) 200
3. (a) 100 (c) 50
(b) 150 (d) 250
4. (a) 6 (c) 10
(b) 9 (d) 5
5. (a) 7 (c) 3
(b) 4
6. (a) 5 r2 (c) 9 r3
(b) 7 r4
7. (a) 30 (d) 50
(b) 25 (e) 10
(c) 20 (f) 100
8. (a) 150 (c) 300
(b) 400
9. (a) $3.50 (d) $1.00
(b) $4.50 (e) $1.50
(c) $8.00 (f) 50¢
10. Teacher check

.... page 27

Set A

1. 10 21. 40
2. 40 22. 50
3. 3 23. 5
4. 25 24. 40
5. 45 25. 35
6. 30 26. 20
7. 7 27. 6
8. 55 28. 50
9. 25 29. 15
10. 10 30. 35
11. 4 31. 9
12. 35 32. 15
13. 30 33. 50
14. 45 34. 25
15. 10 35. 1
16. 5 36. 10
17. 55 37. 0
18. 15 38. 150
19. 2 39. 8
20. 20 40. 30

Set B

1. 50 21. 55
2. 25 22. 15
3. 9 23. 10
4. 10 24. 20
5. 10 25. 30
6. 0 26. 45
7. 1 27. 4
8. 15 28. 5
9. 15 29. 25
10. 35 30. 10
11. 8 31. 7
12. 30 32. 35
13. 35 33. 50
14. 15 34. 30
15. 5 35. 3
16. 50 36. 55
17. 40 37. 20
18. 50 38. 40
19. 2 39. 8
20. 40 40. 25

.... page 28

Set A

1. (a) 42 (c) 54
(b) 30 (d) 24
2. (a) 3 (c) 5
(b) 9 (d) 10
3. (a) 6 (c) 48
(b) 24 (d) 18
4. (a) 60 (c) 36
(b) 18 (d) 54
5. (a) 48 (c) 36
(b) 30
6. 12, 18, 30, 36, 48, 54, 60
7. (a) 9 (c) 4
(b) 6
8. (a) 48 (c) 54
(b) 24
9. (a) ✘ (f) ✔
(b) ✔ (g) ✘
(c) ✘ (h) ✔
(d) ✔ (i) ✔
(e) ✔ (j) ✘
10. Teacher check

Set B

1. (a) 18 (c) 48
(b) 60 (d) 36
2. (a) 6 (c) 1
(b) 7 (d) 4
3. (a) 12 (c) 54
(b) 42 (d) 36
4. (a) 42 (c) 24
(b) 48 (d) 12
5. (a) 18 (c) 54
(b) 42
6. 54, 48, 36, 30, 12, 6
7. (a) 6 (c) 2
(b) 0
8. (a) 60 (c) 30
(b) 18
9. (a) ✔ (f) ✔
(b) ✘ (g) ✘
(c) ✔ (h) ✔
(d) ✔ (i) ✔
(e) ✘ (j) ✘
10. Teacher check

.... page 29

1. (a) 46 (b) 45
2. (a) 24 (c) 60
(b) 360 (d) 600
3. (a) 60 (c) 240
(b) 120 (d) 30
4. (a) 7 (c) 10
(b) 9
5. (a) 5 (c) 3
(b) 8
6. (a) 4 r2 (c) 6 r4
(b) 9 r3
7. (a) 18 (d) 60
(b) 48 (e) 36
(c) 30 (f) 54
8. (a) 120 (c) 540
(b) 360
9. (a) $3.60 (d) $6.40
(b) $5.40 (e) $4.60
(c) $9.00
10. Teacher check

.... page 30

Set A

1. 12 21. 66
2. 42 22. 54
3. 4 23. 3
4. 24 24. 6
5. 42 25. 24
6. 60 26. 360
7. 8 27. 1
8. 12 28. 36
9. 18 29. 48
10. 24 30. 30
11. 2 31. 7
12. 30 32. 48
13. 60 33. 54
14. 48 34. 18
15. 9 35. 5
16. 18 36. 54
17. 36 37. 30
18. 0 38. 36
19. 10 39. 6
20. 42 40. 60

Set B

1. 60 21. 54
2. 18 22. 180
3. 3 23. 2
4. 48 24. 42
5. 36 25. 30
6. 36 26. 48
7. 1 27. 10
8. 60 28. 18
9. 6 29. 18
10. 30 30. 24
11. 7 31. 9
12. 54 32. 30
13. 24 33. 42
14. 0 34. 60
15. 6 35. 4
16. 36 36. 12
17. 48 37. 12
18. 54 38. 42
19. 5 39. 8
20. 6 40. 24

.... page 31

Set A

1. (a) 28 (c) 63
(b) 70 (d) 14
2. (a) 2 (c) 5
(b) 8 (d) 6
3. (a) 63 (c) 7
(b) 28 (d) 35
4. (a) 21 (c) 42
(b) 14 (d) 7
5. (a) 21 (c) 35
(b) 49
6. 14, 21, 28, 49, 56, 70
7. (a) 8 (c) 10
(b) 7
8. (a) 70 (c) 21
(b) 28
9. (a) ✔ (f) ✔
(b) ✘ (g) ✘
(c) ✘ (h) ✔
(d) ✔ (i) ✘
(e) ✔ (j) ✔
10. Teacher check

Set B

1. (a) 21 (c) 49
(b) 35 (d) 70
2. (a) 1 (c) 10
(b) 9 (d) 3
3. (a) 42 (c) 56
(b) 14 (d) 49
4. (a) 28 (c) 56
(b) 14 (d) 49
5. (a) 63 (c) 56
(b) 28
6. 70, 56, 42, 35, 21, 14
7. (a) 4 (c) 6
(b)7
8. (a) 63 (c) 28
(b) 63
9. (a) ✔ (f) ✔
(b) ✘ (g) ✘
(c) ✔ (h) ✔
(d) ✔ (i) ✔
(e) ✘ (j) ✔
10. Teacher check

.... page 32

1. (a) 54 (b) 53
2. (a) 420 (c) 70
(b) 560 (d) 42
3. (a) 350 (c) 35
(b) 140 (d) 7
4. (a) 6 (c) 9
(b) 5
5. (a) 8 (c) 3
(b) 6
6. (a) 5 r2 (c) 3 r4
(b) 9 r1
7. (a) 28 (d) 63
(b) 35 (e) 21
(c) 49 (f) 42
8. (a) 210 (c) 700
(b) 420
9. (a) $3.50 (d) $6.50
(b) $5.60 (e) $4.40
(c) $2.10
10. Teacher check

.... page 33

Set A

1. 21 21. 28
2. 42 22. 14
3. 3 23. 4
4. 7 24. 42
5. 35 25. 49
6. 21 26. 35
7. 7 27. 6
8. 63 28. 70
9. 42 29. 63
10. 49 30. 63
11. 2 31. 10
12. 14 32. 35
13. 56 33. 490
14. 0 34. 70
15. 5 35. 1
16. 28 36. 56
17. 77 37. 70
18. 490 38. 28
19. 9 39. 8
20. 21 40. 0

Set B

1. 14 21. 63
2. 56 22. 210
3. 6 23. 4
4. 63 24. 28
5. 28 25. 56
6. 35 26. 28
7. 1 27. 7
8. 21 28. 42
9. 70 29. 42
10. 21 30. 42
11. 5 31. 8
12. 7 32. 56
13. 0 33. 35
14. 63 34. 280
15. 2 35. 3
16. 140 36. 49
17. 49 37. 21
18. 49 38. 70
19. 0 39. 8
20. 14 40. 35

.... page 34

Set A

1. (a) 56 (c) 32
(b) 48 (d) 72
2. (a) 4 (c) 10
(b) 5 (d) 2
3. (a) 64 (c) 40
(b) 56 (d) 8
4. (a) 64 (c) 32
(b) 56 (d) 8
5. (a) 48 (c) 64
(b) 24
6. 0, 16, 24, 40, 48, 64, 72
7. (a) 6 (c) 9

www.worldteacherspress.com ©World Teachers Press®

Answers

(b) 8
8. (a) 40 (c) 72
(b) 80
9. (a) ✔ (f) ✔
(b) ✘ (g) ✘
(c) ✔ (h) ✔
(d) ✘ (i) ✔
(e) ✔ (j) ✘
10. Teacher check

Set B

1. (a) 24 (c) 64
(b) 80 (d) 40
2. (a) 7 (c) 1
(b) 8 (d) 3
3. (a) 72 (c) 16
(b) 48 (d) 32
4. (a) 16 (c) 24
(b) 48 (d) 80
5. (a) 56 (c) 72
(b) 40
6. 72, 64, 48, 32, 24, 16, 0
7. (a) 3 (c) 4
(b) 8
8. (a) 64 (c) 56
(b) 40
9. (a) ✘ (f) ✔
(b) ✔ (g) ✔
(c) ✔ (h) ✘
(d) ✔ (i) ✔
(e) ✘ (j) ✔
10. Teacher check

.... **page 35**

1. (a) 60 (b) 58
2. (a) 320 (c) 160
(b) 240 (d) 16
3. (a) 32 (c) 8
(b) 24 (d) 40
4. (a) 5 (c) 7
(b) 10
5. (a) 4 (c) 2
(b) 8
6. (a) 9 r2 (c) 2 r5
(b) 4 r3
7. (a) 24 (d) 32
(b) 48 (e) 40
(c) 72 (f) 80
8. (a) 400 (c) 560
(b) 720
9. (a) $3.20 (d) $6.80
(b) $4.80 (e) $5.20
(c) $8.00
10. Teacher check

.... **page 36**

Set A

1. 24 21. 32
2. 48 22. 16
3. 3 23. 4
4. 8 24. 48
5. 40 25. 56
6. 24 26. 40
7. 7 27. 6
8. 72 28. 80
9. 48 29. 72
10. 64 30. 72
11. 2 31. 10
12. 32 32. 40
13. 64 33. 240
14. 0 34. 80
15. 5 35. 1
16. 32 36. 64
17. 88 37. 80
18. 640 38. 32
19. 8 39. 9
20. 24 40. 160

Set B

1. 88 21. 80
2. 64 22. 0
3. 6 23. 4
4. 72 24. 32
5. 32 25. 64
6. 40 26. 32
7. 1 27. 7
8. 24 28. 48
9. 80 29. 48
10. 24 30. 48
11. 5 31. 8
12. 8 32. 64
13. 16 33. 40
14. 72 34. 400
15. 2 35. 3
16. 80 36. 56
17. 56 37. 24
18. 56 38. 80
19. 160 39. 9
20. 16 40. 40

.... **page 37**

Set A

1. (a) 36 (c) 81
(b) 54 (d) 27
2. (a) 10 (c) 4
(b) 7 (d) 9
3. (a) 9 (c) 27
(b) 72 (d) 45
4. (a) 81 (c) 54
(b) 18 (d) 27
5. (a) 45 (c) 72
(b) 27
6. 18, 27, 45, 54, 72, 81
7. (a) 5 (c) 8
(b) 9
8. (a) 36 (c) 45
(b) 54
9. (a) ✔ (f) ✘
(b) ✔ (g) ✔
(c) ✔ (h) ✔
(d) ✘ (i) ✔
(e) ✔ (j) ✘
10. Teacher check

Set B

1. (a) 63 (c) 90
(b) 18 (d) 72
2. (a) 2 (c) 6
(b) 8 (d) 3
3. (a) 54 (c) 81
(b) 18 (d) 36
4. (a) 36 (c) 63
(b) 72 (d) 90
5. (a) 63 (c) 81
(b) 36
6. 90, 81, 63, 54, 36, 27, 9, 0
7. (a) 4 (c) 9
(b) 0
8. (a) 90 (c) 9
(b) 81
9. (a) ✘ (f) ✔
(b) ✘ (g) ✘
(c) ✘ (h) ✔
(d) ✔ (i) ✘
(e) ✔ (j) ✔
10. Teacher check

.... **page 38**

1. (a) 66 (b) 68
2. (a) 360 (c) 180
(b) 900 (d) 36
3. (a) 90 (c) 9
(b) 27 (d) 45
4. (a) 5 (c) 8
(b) 10
5. (a) 3 (c) 9
(b) 6
6. (a) 4 r2 (c) 8 r7
(b) 6 r5
7. (a) 36 (d) 45
(b) 81 (e) 18
(c) 54 (f) 63
8. (a) 630 (c) 270
(b) 810
9. (a) $7.20 (d) $2.80
(b) $3.60 (e) $6.40
(c) $3.60
10. Teacher check

.... **page 39**

Set A

1. 27 21. 36
2. 54 22. 18
3. 3 23. 4
4. 9 24. 54
5. 45 25. 63
6. 27 26. 45
7. 7 27. 6
8. 36 28. 90
9. 54 29. 81
10. 63 30. 81
11. 2 31. 10
12. 18 32. 72
13. 72 33. 810
14. 0 34. 90
15. 5 35. 1
16. 81 36. 45
17. 99 37. 90
18. 9 38. 36
19. 9 39. 8
20. 27 40. 360

Set B

1. 36 21. 54
2. 90 22. 0
3. 9 23. 3
4. 72 24. 27
5. 63 25. 180
6. 36 26. 270
7. 5 27. 7
8. 45 28. 36
9. 81 29. 90
10. 81 30. 63
11. 9 31. 10
12. 0 32. 9
13. 27 33. 72
14. 45 34. 27
15. 4 35. 8
16. 18 36. 54
17. 45 37. 99
18. 18 38. 54
19. 6 39. 1
20. 81 40. 90

.... **page 40**

Set A

1. (a) 50 (c) 30
(b) 60 (d) 100
2. (a) 3 (c) 8
(b) 4 (d) 11
3. (a) 40 (c) 80
(b) 10 (d) 30
4. (a) 60 (c) 50
(b) 30 (d) 90
5. (a) 40 (c) 80
(b) 60
6. 10, 20, 30, 50, 60, 80, 90
7. (a) 7 (c) 5
(b) 10
8. (a) 110 (c) 90
(b) 70
9. (a) ✔ (f) ✘
(b) ✔ (g) ✘
(c) ✘ (h) ✔
(d) ✔ (i) ✔
(e) ✔ (j) ✘
10. Teacher check

Set B

1. (a) 20 (c) 90
(b) 80 (d) 120
2. (a) 5 (c) 9
(b) 6 (d) 20
3. (a) 70 (c) 50
(b) 100 (d) 90
4. (a) 20 (c) 30
(b) 70 (d) 100
5. (a) 30 (c) 90
(b)70
6. 90, 80, 60, 40, 30, 20, 0
7. (a) 2 (c) 6
(b) 10
8. (a) 50 (c) 40
(b) 60
9. (a) ✔ (f) ✘
(b) ✘ (g) ✔
(c) ✘ (h) ✘
(d) ✔ (i) ✔
(e) ✔ (j) ✘
10. Teacher check

.... **page 41**

1. (a) 74 (b) 71
2. (a) 1000 (c) 80
(b) 600 (d) 40
3. (a) 10 (c) 400
(b) 30 (d) 50
4. (a) 15 (c) 12
(b) 27 (d) 60
5. (a) 8 (c) 2
(b) 10
6. (a) 7 r2 (c) 10 r3
(b) 6 r7
7. (a) 100 (d) 150
(b) 40 (e) 20
(c) 70 (f) 200
8. (a) 600 (c) 2000
(b) 1000
9. (a) $8.00 (d) $12.00
(b) $4.00 (e) $16.00
(c) $12.00
10. Teacher check

.... **page 42**

Set A

1. 40 21. 170
2. 80 22. 80
3. 7 23. 15
4. 490 24. 100
5. 100 25. 30
6. 100 26. 20
7. 4 27. 20
8. 790 28. 110
9. 60 29. 110
10. 60 30. 60
11. 10 31. 4
12. 90 32. 10
13. 130 33. 210
14. 0 34. 0
15. 5 35. 7
16. 290 36. 30
17. 50 37. 70
18. 30 38. 90
19. 30 39. 6
20. 890 40. 140

Set B

1. 90 21. 160
2. 40 22. 500
3. 10 23. 20
4. 40 24. 140
5. 50 25. 30
6. 90 26. 300
7. 15 27. 60
8. 80 28. 190
9. 80 29. 150
10. 100 30. 0
11. 70 31. 80
12. 0 32. 160
13. 100 33. 210
14. 700 34. 400
15. 40 35. 50
16. 90 36. 290
17. 110 37. 200
18. 200 38. 90
19. 600 39. 30
20. 110 40. 240

.... **page 43**

Set A

1. (a) 48 (c) 72
(b) 108 (d) 36
2. (a) 2 (c) 6
(b) 9 (d) 7
3. (a) 36 (c) 96
(b) 12 (d) 60
4. (a) 36 (c) 84
(b) 60 (d) 96
5. (a) 36 (c) 60
(b) 84
6. 12, 36, 60, 72, 96, 108
7. (a) 2 (c) 6
(b) 10
8. (a) 84 (c) 60
(b) 108
9. (a) ✔ (f) ✔
(b) ✘ (g) ✔

©World Teachers Press® www.worldteacherspress.com

(c) ✔ (h) ✘
(d) ✘ (i) ✔
(e) ✘ (j) ✔
10. Teacher check

Set B

1. (a) 24 (c) 60
(b) 96 (d) 120
2. (a) 5 (c) 3
(b) 10 (d) 4
3. (a) 48 (c) 108
(b) 72 (d) 24
4. (a) 96 (c) 24
(b) 48, 36 (d) 60
5. (a) 48 (c) 108
(b) 72
6. 108, 96, 72, 48, 36, 24, 0
7. (a) 9 (c) 8
(b) 0
8. (a) 96 (c) 84
(b) 48
9. (a) ✔ (f) ✔
(b) ✘ (g) ✘
(c) ✔ (h) ✔
(d) ✘ (i) ✔
(e) ✔ (j) ✘
10. Teacher check

.... page 44

1. (a) 87 (b) 89
2. (a) 120 (c) 80
(b) 96 (d) 40
3. (a) 12 (c) 10
(b) 60 (d) 30
4. (a) 5 (c) 8
(b) 10
5. (a) 3 (c) 10
(b) 5
6. (a) 4 r2 (c) 2 r3
(b) 3 r5
7. (a) 24 (d) 84
(b) 120 (e) 48
(c) 60 (f) 96
8. (a) 360 (c) 1080
(b) 840
9. (a) $1.20 (d) $3.80
(b) 60¢ (e) $4.40
(c) 60¢
10. Teacher check

.... page 45

Set A

1. 36	21. 48
2. 72	22. 24
3. 3	23. 4
4. 12	24. 72
5. 60	25. 84
6. 36	26. 60
7. 7	27. 6
8. 48	28. 0
9. 72	29. 108
10. 0	30. 36
11. 2	31. 10
12. 24	32. 96
13. 96	33. 12
14. 84	34. 120
15. 5	35. 1
16. 108	36. 60
17. 132	37. 120
18. 12	38. 48
19. 9	39. 8
20. 36	40. 0

Set B

1. 48	21. 36
2. 120	22. 120
3. 8	23. 1
4. 96	24. 108
5. 84	25. 60
6. 48	26. 108
7. 6	27. 7
8. 60	28. 24
9. 108	29. 72
10. 0	30. 84
11. 2	31. 3
12. 0	32. 48
13. 24	33. 96
14. 60	34. 36
15. 4	35. 5
16. 72	36. 12
17. 120	37. 132
18. 24	38. 72
19. 9	39. 0
20. 36	40. 84

.... page 46

Set A

1. (a) 90 (c) 120
(b) 60 (d) 135
2. (a) 8 (c) 1
(b) 5 (d) 7
3. (a) 135 (c) 30
(b) 105 (d) 45
4. (a) 150 (c) 90
(b) 30 (d) 15
5. (a) 45 (c) 120
(b) 75
6. 30, 45, 75, 90, 120, 135
7. (a) 3 (c) 6
(b) 4
8. (a) 60 (c) 75
(b) 120
9. (a) ✘ (f) ✔
(b) ✔ (g) ✔
(c) ✔ (h) ✘
(d) ✘ (i) ✔
(e) ✔ (j) ✔
10. Teacher check

Set B

1. (a) 150 (c) 105
(b) 75 (d) 30
2. (a) 10 (c) 2
(b) 6 (d) 3
3. (a) 60 (c) 75
(b) 120 (d) 15
4. (a) 105 (c) 30
(b) 45 (d) 135
5. (a) 60 (c) 135
(b) 90
6. 135, 120, 90, 60, 45, 30, 0
7. (a) 10 (c) 2
(b) 5
8. (a) 90 (c) 105
(b) 45
9. (a) ✔ (f) ✔
(b) ✔ (g) ✔
(c) ✔ (h) ✔
(d) ✘ (i) ✘
(e) ✘ (j) ✔
10. Teacher check

.... page 47

1. (a) 108 (b) 110
2. (a) 120 (c) 30
(b) 90 (d) 150
3. (a) 60 (c) 30
(b) 75 (d) 15
4. (a) 3 (c) 9
(b) 8 (d) 10
5. (a) 4 (c) 7
(b) 10
6. (a) 3 r3 (c) 9 r5
(b) 7 r2
7. (a) 60 (d) 45
(b) 105 (e) 30
(c) 75 (f) 135
8. (a) 450 (c) 1350
(b) 900
9. (a) $4.50 (d) $5.50
(b) $10.50 (e) $9.50
(c) $15.00
10. Teacher check

.... page 48

Set A

1. 45	21. 135
2. 90	22. 75
3. 3	23. 4
4. 15	24. 90
5. 75	25. 105
6. 45	26. 120
7. 7	27. 6
8. 60	28. 75
9. 90	29. 0
10. 150	30. 135
11. 2	31. 1
12. 30	32. 135
13. 60	33. 150
14. 30	34. 15
15. 5	35. 8
16. 0	36. 105
17. 30	37. 120
18. 60	38. 105
19. 9	39. 10
20. 45	40. 120

Set B

1. 135	21. 30
2. 15	22. 60
3. 4	23. 9
4. 45	24. 60
5. 105	25. 45
6. 105	26. 30
7. 6	27. 5
8. 30	28. 105
9. 0	29. 75
10. 75	30. 150
11. 1	31. 2
12. 0	32. 135
13. 150	33. 60
14. 135	34. 45
15. 8	35. 7
16. 15	36. 120
17. 120	37. 90
18. 120	38. 90
19. 10	39. 3
20. 75	40. 90

.... page 49

Set A

1. (a) 100 (c) 140
(b) 180 (d) 80
2. (a) 3 (c) 5
(b) 8 (d) 6
3. (a) 20 (c) 160
(b) 80 (d) 60
4. (a) 40 (c) 140
(b) 60 (d) 20
5. (a) 80 (c) 160
(b) 100
6. 40, 60, 100, 120, 160, 180
7. (a) 5 (c) 0
(b) 9
8. (a) 160 (c) 180
(b) 100
9. (a) ✔ (f) ✔
(b) ✔ (g) ✘
(c) ✘ (h) ✘
(d) ✘ (i) ✔
(e) ✔ (j) ✔
10. Teacher check

Set B

1. (a) 60 (c) 160
(b) 40 (d) 200
2. (a) 4 (c) 10
(b) 9 (d) 2
3. (a) 120 (c) 180
(b) 40 (d) 140
4. (a) 120 (c) 160
(b) 100 (d) 200
5. (a) 140 (c) 180
(b) 60
6. 180, 160, 120, 80, 60, 40, 0
7. (a) 6 (c) 3
(b) 10
8. (a) 200 (c) 140
(b) 80
9. (a) ✔ (f) ✘
(b) ✔ (g) ✔
(c) ✘ (h) ✔
(d) ✔ (i) ✔
(e) ✔ (j) ✔
10. Teacher check

.... page 50

1. (a) 145 (b) 142
2. (a) 160 (c) 40
(b) 120 (d) 200
3. (a) 300 (c) 40
(b) 80 (d) 20
4. (a) 5 (c) 10
(b) 3 (d) 7
5. (a) 4 (c) 10
(b) 7
6. (a) 3 r8 (c) 8 r15
(b) 5 r5
7. (a) 40 (d) 160
(b) 140 (e) 200
(c) 100 (f) 400
8. (a) 600 (c) 2000
(b) 1400
9. (a) $12.00 (d) $8.00
(b) $18.00 (e) $2.00
(c) $6.00
10. Teacher check

.... page 51

Set A

1. 100	21. 140
2. 60	22. 40
3. 5	23. 4
4. 60	24. 140
5. 120	25. 240
6. 200	26. 100
7. 3	27. 6
8. 120	28. 40
9. 180	29. 200
10. 80	30. 160
11. 8	31. 9
12. 80	32. 20
13. 80	33. 0
14. 180	34. 120
15. 10	35. 7
16. 100	36. 160
17. 60	37. 160
18. 140	38. 0
19. 2	39. 1
20. 180	40. 200

Set B

1. 200	21. 160
2. 120	22. 60
3. 7	23. 8
4. 40	24. 80
5. 140	25. 180
6. 80	26. 100
7. 5	27. 3
8. 100	28. 60
9. 200	29. 120
10. 40	30. 160
11. 20	31. 9
12. 220	32. 140
13. 100	33. 60
14. 140	34. 200
15. 4	35. 6
16. 160	36. 120
17. 80	37. 220
18. 180	38. 0
19. 10	39. 1
20. 20	40. 180

.... page 52

Set A

1. (a) 150 (c) 100
(b) 250 (d) 125
2. (a) 4 (c) 10
(b) 3 (d) 7
3. (a) 150 (c) 75
(b) 225 (d) 125
4. (a) 175 (c) 200
(b) 75 (d) 25
5. (a) 100 (c) 200
(b) 125
6. 25, 75, 125, 150, 200, 225
7. (a) 8 (c) 9
(b) 6
8. (a) 200 (c) 50
(b) 125
9. (a) ✔ (f) ✘
(b) ✘ (g) ✔
(c) ✔ (h) ✔
(d) ✔ (i) ✘
(e) ✔ (j) ✔
10. Teacher check

www.worldteacherspress.com ©World Teachers Press®

Answers

Set B

1. (a) 175 (c) 75
(b) 200 (d) 225
2. (a) 8 (c) 6
(b) 9 (d) 2
3. (a) 25 (c) 175
(b) 50 (d) 100
4. (a) 75 (c) 150
(b) 250 (d) 100
5. (a) 175 (c) 225
(b) 150
6. 225, 200, 150, 100, 75, 50, 0
7. (a) 4 (c) 7
(b) 10
8. (a) 75 (c) 175
(b) 100
9. (a) ✘ (f) ✔
(b) ✔ (g) ✘
(c) ✔ (h) ✔
(d) ✘ (i) ✔
(e) ✘ (j) ✔
10. Teacher check

.... page 53

1. (a) 178 (b) 180
2. (a) 250 (c) 100
(b) 150 (d) 50
3. (a) 75 (c) 100
(b) 25 (d) 50
4. (a) 5 (c) 4
(b) 9 (d) 6
5. (a) 10 (c) 5
(b) 7
6. (a) 3 r4 (c) 8 r15
(b) 5 r5
7. (a) 75 (d) 175
(b) 100 (e) 125
(c) 225 (f) 200
8. (a) 1000 (c) 2500
(b) 1500
9. (a) $7.50 (d) $2.50
(b) $17.50 (e) $2.50
(c) $25.00
10. Teacher check

.... page 54

Set A

1. 75	21. 150
2. 150	22. 200
3. 4	23. 2
4. 225	24. 50
5. 125	25. 200
6. 250	26. 50
7. 8	27. 5
8. 175	28. 0
9. 175	29. 250
10. 75	30. 125
11. 1	31. 7
12. 200	32. 125
13. 50	33. 225
14. 100	34. 225
15. 3	35. 10
16. 150	36. 75
17. 100	37. 500
18. 175	38. 0
19. 6	39. 9
20. 100	40. 25

Set B

1. 1000	21. 150
2. 0	22. 175
3. 2	23. 4
4. 75	24. 100
5. 75	25. 100
6. 225	26. 100
7. 5	27. 8
8. 25	28. 150
9. 225	29. 50
10. 125	30. 75
11. 7	31. 1
12. 125	32. 200
13. 250	33. 175
14. 50	34. 250
15. 10	35. 3
16. 0	36. 175
17. 200	37. 125
18. 200	38. 150
19. 9	39. 6
20. 50	40. 225

.... page 55

Set A

1. (a) 200 (c) 400
(b) 100 (d) 150
2. (a) 4 (c) 7
(b) 2 (d) 8
3. (a) 450 (c) 100
(b) 200 (d) 150
4. (a) 100 (c) 350
(b) 500 (d) 150
5. (a) 200 (c) 400
(b) 250
6. 100, 150, 250, 300, 400, 450
7. (a) 7 (c) 9
(b) 5
8. (a) 350 (c) 100
(b) 500
9. (a) ✔ (f) ✔
(b) ✘ (g) ✘
(c) ✔ (h) ✔
(d) ✔ (i) ✘
(e) ✔ (j) ✔
10. Teacher check

Set B

1. (a) 500 (c) 350
(b) 300 (d) 250
2. (a) 6 (c) 9
(b) 10 (d) 5
3. (a) 400 (c) 50
(b) 250 (d) 500
4. (a) 400 (c) 150
(b) 300 (d) 500
5. (a) 350 (c) 450
(b) 300
6. 500, 400, 350, 250, 200, 150, 50, 0
7. (a) 10 (c) 8
(b) 4
8. (a) 250 (c) 300
(b) 450
9. (a) ✔ (f) ✘
(b) ✔ (g) ✔
(c) ✘ (h) ✘
(d) ✘ (i) ✔
(e) ✔ (j) ✔
10. Teacher check

.... page 56

1. (a) 354 (b) 351
2. (a) 300 (c) 100
(b) 400 (d) 800
3. (a) 250 (c) 50
(b) 150 (d) 100
4. (a) 5 (c) 3
(b) 9 (d) 6
5. (a) 3 (c) 9
(b) 6
6. (a) 2 r20 (c) 9 r10
(b) 7 r30
7. (a) 150 (d) 350
(b) 250 (e) 500
(c) 400 (f) 450
8. (a) 3000 (c) 3500
(b) 5000
9. (a) $25.00 (d) $25.00
(b) $45.00 (e) $5.00
(c) $20.00
10. Teacher check

.... page 57

Set A

1. 150	21. 300
2. 300	22. 400
3. 4	23. 2
4. 450	24. 100
5. 250	25. 400
6. 500	26. 0
7. 8	27. 5
8. 350	28. 0
9. 350	29. 500
10. 150	30. 250
11. 1	31. 7
12. 400	32. 250
13. 100	33. 450
14. 200	34. 450
15. 3	35. 10
16. 300	36. 150
17. 200	37. 0
18. 350	38. 400
19. 6	39. 9
20. 200	40. 50

Set B

1. 500	21. 2000
2. 0	22. 350
3. 2	23. 4
4. 150	24. 200
5. 150	25. 200
6. 450	26. 200
7. 5	27. 8
8. 50	28. 300
9. 450	29. 100
10. 250	30. 250
11. 7	31. 1
12. 250	32. 400
13. 500	33. 350
14. 100	34. 500
15. 0	35. 3
16. 0	36. 350
17. 400	37. 250
18. 400	38. 300
19. 9	39. 6
20. 100	40. 450

Review answers

.... page 58

Set A

1. 14	21. 16
2. 0	22. 9
3. 8	23. 4 r1
4. 7	24. 5 r2
5. 6	25. 80
6. 12	26. 180
7. 8	27. 8
8. 12	28. 7
9. 120	29. 10
10. 120	30. 0
11. 40	31. 5
12. 60	32. 3
13. 6	33. 9
14. 6	34. 4
15. 12	35. 12
16. 30	36. 15
17. 22	37. 10
18. 27	38. 10
19. 12	39. 40
20. 15	40. 90

Set B

1. 8	21. 18
2. 27	22. 27
3. 9	23. 3 r1
4. 9	24. 7 r1
5. 8	25. 100
6. 9	26. 210
7. 16	27. 0
8. 9	28. 8
9. 80	29. 9
10. 210	30. 9
11. 80	31. 4
12. 120	32. 15
13. 8	33. 10
14. 8	34. 6
15. 16	35. 8
16. 15	36. 21
17. 20	37. 7
18. 21	38. 1
19. 8	39. 140
20. 21	40. 180

.... page 59

Set A

1. 0	21. 8
2. 25	22. 40
3. 4	23. 3 r1
4. 6	24. 7 r2
5. 12	25. 160
6. 20	26. 250
7. 8	27. 24
8. 35	28. 6
9. 280	29. 9
10. 400	30. 10
11. 16	31. 4
12. 20	32. 5
13. 3	33. 5
14. 10	34. 7
15. 12	35. 16
16. 35	36. 45
17. 36	37. 10
18. 50	38. 4
19. 0	39. 280
20. 30	40. 300

Set B

1. 36	21. 24
2. 0	22. 10
3. 4 r2	23. 5 r3
4. 5 r2	24. 8 r2
5. 16	25. 200
6. 15	26. 350
7. 24	27. 7
8. 5	28. 45
9. 240	29. 4
10. 300	30. 0
11. 80	31. 20
12. 100	32. 25
13. 5	33. 7
14. 7	34. 4
15. 16	35. 12
16. 40	36. 25
17. 40	37. 10
18. 55	38. 5
19. 0	39. 400
20. 20	40. 150

.... page 60

Set A

1. 18	21. 6
2. 56	22. 63
3. 6	23. 2 r2
4. 6	24. 4 r2
5. 18	25. 420
6. 28	26. 630
7. 12	27. 4
8. 42	28. 5
9. 360	29. 0
10. 210	30. 2
11. 12	31. 30
12. 14	32. 35
13. 5	33. 4
14. 9	34. 7
15. 18	35. 30
16. 56	36. 49
17. 54	37. 2
18. 21	38. 10
19. 24	39. 480
20. 21	40. 350

Set B

1. 30	21. 0
2. 0	22. 49
3. 7	23. 3 r2
4. 8	24. 7 r2
5. 24	25. 300
6. 21	26. 490
7. 42	27. 10
8. 7	28. 0
9. 180	29. 6
10. 560	30. 9
11. 120	31. 6
12. 140	32. 14
13. 3	33. 3
14. 4	34. 4
15. 24	35. 6
16. 42	36. 63
17. 30	37. 8
18. 28	38. 1
19. 12	39. 240
20. 28	40. 420

Answers

.... *page 61*

Set A

1. 40	*21. 72*
2. 81	*22. 81*
3. 4	*23. 2 r3*
4. 2	*24. 3 r4*
5. 32	*25. 640*
6. 36	*26. 810*
7. 16	*27. 4*
8. 63	*28. 5*
9. 560	*29. 7*
10. 270	*30. 8*
11. 16	*31. 40*
12. 18	*32. 45*
13. 5	*33. 5*
14. 1	*34. 7*
15. 32	*35. 8*
16. 27	*36. 36*
17. 56	*37. 3*
18. 54	*38. 4*
19. 8	*39. 320*
20. 27	*40. 900*

Set B

1. 48	*21. 56*
2. 0	*22. 9*
3. 2	*23. 5 r4*
4. 10	*24. 9 r5*
5. 24	*25. 800*
6. 27	*26. 540*
7. 24	*27. 10*
8. 72	*28. 7*
9. 480	*29. 2*
10. 450	*30. 10*
11. 160	*31. 20*
12. 180	*32. 9*
13. 3	*33. 1*
14. 9	*34. 8*
15. 72	*35. 32*
16. 90	*36. 81*
17. 24	*37. 4*
18. 18	*38. 5*
19. 72	*39. 800*
20. 45	*40. 630*

.... *page 62*

Set A

1. 70	*21. 10*
2. 0	*22. 60*
3. 8	*23. 8 r7*
4. 6	*24. 3 r3*
5. 30	*25. 1000*
6. 12	*26. 360*
7. 50	*27. 5*
8. 12	*28. 4*
9. 400	*29. 0*
10. 240	*30. 8*
11. 20	*31. 5*
12. 24	*32. 6*
13. 3	*33. 9*
14. 9	*34. 2*
15. 30	*35. 90*
16. 48	*36. 48*
17. 60	*37. 4*
18. 36	*38. 10*
19. 60	*39. 600*
20. 36	*40. 1200*

Set B

1. 0	*21. 80*
2. 96	*22. 48*
3. 10	*23. 9 r4*
4. 7	*24. 4 r2*
5. 10	*25. 200*
6. 36	*26. 600*
7. 10	*27. 7*
8. 24	*28. 0*
9. 500	*29. 10*
10. 0	*30. 10*
11. 200	*31. 50*
12. 240	*32. 60*
13. 6	*33. 4*
14. 5	*34. 4*
15. 50	*35. 60*
16. 84	*36. 60*
17. 90	*37. 6*
18. 72	*38. 4*
19. 90	*39. 500*
20. 0	*40. 480*

www.worldteacherspress.com ©World Teachers Press®

NOTES

NOTES

www.worldteacherspress.com ©World Teachers Press®